RADIATIVE FORCING OF CLIMATE CHANGE

EXPANDING THE CONCEPT AND ADDRESSING UNCERTAINTIES

Committee on Radiative Forcing Effects on Climate

Climate Research Committee

Board on Atmospheric Sciences and Climate

Division on Earth and Life Studies

NATIONAL RESEARCH COUNCIL
OF THE NATIONAL ACADEMIES

THE NATIONAL ACADEMIES PRESS
Washington, D.C.
www.nap.edu

THE NATIONAL ACADEMIES PRESS   500 Fifth Street, NW Washington, DC 20001

NOTICE: The project that is the subject of this report was approved by the Governing Board of the National Research Council, whose members are drawn from the councils of the National Academy of Sciences, the National Academy of Engineering, and the Institute of Medicine. The members of the committee responsible for the report were chosen for their special competences and with regard for appropriate balance.

Support for this project was provided by the National Oceanic and Atmospheric Administration under Contract No. NASW-01008. Any opinions, findings, and conclusions, or recommendations expressed in this publication are those of the author(s) and do not necessarily reflect the views of the organizations or agencies that provided support for the project.

International Standard Book Number 0-309-09506-9 (Book)
International Standard Book Number 0-309-54688-5 (PDF)
Library of Congress Control Number 2005921687

Additional copies of this report are available from the National Academies Press, 500 Fifth Street, N.W., Lockbox 285, Washington, DC 20055; (800) 624-6242 or (202) 334-3313 (in the Washington metropolitan area); Internet, http://www.nap.edu.

Cover: Images obtained from the Clouds and Earth's Radiant Energy System (CERES) instrument on board NASA's Aqua satellite on June 22, 2002. The image on the front cover shows the amount of infrared energy, or heat, emitted by the Earth and its atmosphere to space. Clear, warm land regions (shown in yellow) emit the most heat. High, cold clouds (shown in blue and white) emit less heat to space. The image on the back cover shows the amount of sunlight reflected back to space. Clear ocean areas (shown in dark blue) reflect the least amount of sunlight back to space, and clouds and snow-covered surfaces (shown in white and green) reflect the most sunlight back to space. SOURCE: National Aeronautics and Space Administration (NASA) Goddard Space Flight Center.

THE NATIONAL ACADEMIES
Advisers to the Nation on Science, Engineering, and Medicine

The **National Academy of Sciences** is a private, nonprofit, self-perpetuating society of distinguished scholars engaged in scientific and engineering research, dedicated to the furtherance of science and technology and to their use for the general welfare. Upon the authority of the charter granted to it by the Congress in 1863, the Academy has a mandate that requires it to advise the federal government on scientific and technical matters. Dr. Bruce M. Alberts is president of the National Academy of Sciences.

The **National Academy of Engineering** was established in 1964, under the charter of the National Academy of Sciences, as a parallel organization of outstanding engineers. It is autonomous in its administration and in the selection of its members, sharing with the National Academy of Sciences the responsibility for advising the federal government. The National Academy of Engineering also sponsors engineering programs aimed at meeting national needs, encourages education and research, and recognizes the superior achievements of engineers. Dr. Wm. A. Wulf is president of the National Academy of Engineering.

The **Institute of Medicine** was established in 1970 by the National Academy of Sciences to secure the services of eminent members of appropriate professions in the examination of policy matters pertaining to the health of the public. The Institute acts under the responsibility given to the National Academy of Sciences by its congressional charter to be an adviser to the federal government and, upon its own initiative, to identify issues of medical care, research, and education. Dr. Harvey V. Fineberg is president of the Institute of Medicine.

The **National Research Council** was organized by the National Academy of Sciences in 1916 to associate the broad community of science and technology with the Academy's purposes of furthering knowledge and advising the federal government. Functioning in accordance with general policies determined by the Academy, the Council has become the principal operating agency of both the National Academy of Sciences and the National Academy of Engineering in providing services to the government, the public, and the scientific and engineering communities. The Council is administered jointly by both Academies and the Institute of Medicine. Dr. Bruce M. Alberts and Dr. Wm. A. Wulf are chair and vice chair, respectively, of the National Research Council.

www.national-academies.org

CLIMATE RESEARCH COMMITTEE

Preface

Radiative forcing is a way to quantify an energy imbalance imposed on the climate system either externally (e.g., solar energy output or volcanic emissions) or by human activities (e.g., deliberate land modification or emissions of greenhouse gases, aerosols, and their precursors). The concept of radiative forcing has been central for guiding climate research and policy over the past two decades. There are several reasons for this. It provides a simple yet fundamental index that allows us to look at how climate change is driven by the energy imbalance of the Earth system. It is successful in predicting change in global mean surface temperature as computed from climate models and it, thus, allows quantitative comparison of the contributions of different agents to climate change. It is easy to compute and is reproducible across models and therefore offers a convenient common metric on which policy research and recommendations can be based.

New studies on climate forcing agents not conventionally considered have, however, raised doubts as to the continued viability of the radiative forcing concept. For example, the climatic effects from light-absorbing aerosols or land-use changes do not lend themselves to quantification using the traditional radiative forcing concept. Aerosol effects on clouds are difficult to describe in terms of simple radiative forcing. These challenges have raised the question of whether the radiative forcing concept has outlived its usefulness and, if so, what new climate change metrics should be used.

To address these issues, the U.S. Climate Change Science Program (CCSP) asked the National Academies to undertake a study to evaluate the current state of knowledge on radiative forcings and to identify relevant

---

BOX P-1
**Statement of Task for the Committee on Radiative Forcing
Effects on Climate**

This study will examine the current state of knowledge regarding the direct and indirect radiative forcing effects of gases, aerosols, land use, and solar variability on the climate of the Earth's surface and atmosphere and it will identify research needed to improve our understanding of these effects. Specifically, this study will:

1. Summarize what is known about the direct and indirect radiative effects caused by individual forcing agents, including the spatial and temporal scales over which specific forcing agents may be important;
2. Evaluate techniques (e.g., modeling, laboratory, observations, and field experiments) used to estimate direct and indirect radiative effects of specific forcing agents;
3. Identify key gaps in the understanding of radiative forcing effects on climate;
4. Identify key uncertainties in projections of future radiative forcing effects on climate;
5. Recommend near- and longer-term research priorities for improving our understanding and projections of radiative forcing effects on climate.

---

research needs. In response, the Committee on Radiative Forcing Effects on Climate was formed. The committee was charged to examine the current state of knowledge of how gases, aerosols, land use, and solar variability force the climate system, identify key gaps in understanding, and recommend research priorities (see Box P-1 for the full statement of task). This report presents the committee's findings and recommendations.

The committee began its discussions with a good dose of skepticism about the continued viability of the radiative forcing concept. In the end, however, one of our major findings is that the concept retains considerable value. It needs to be expanded to account for the vertical and regional structure of radiative forcing and also for nonradiative climate forcings. We propose several new research avenues that should be pursued to accomplish this expansion. We present an agenda for addressing uncertainties in forcings and climate effects from conventional and nonconventional agents. We make specific recommendations for using past climate records to improve our understanding of the relationship of radiative forcing to climate change and for developing an observational strategy aimed at continuous monitoring of climate forcing variables for the indefinite future. Finally, we examine ways to improve the application of radiative and nonradiative forcing metrics in policy analyses directed at climate change.

The committee met four times over the course of a year to gather information and to deliberate over findings and recommendations. We thank the following speakers who shared their knowledge with the committee: James Anderson, Harvard University; Theodore L. Anderson, University of Washington; Gordon Bonan, National Center for Atmospheric Research; Thomas Crowley, Duke University; Kea Duckenfield, National Oceanic and Atmospheric Administration (NOAA); Jerry Elwood, Department of Energy; David Fahey, NOAA Aeronomy Laboratory; Jay Fein, National Science Foundation; Peter Gent, National Center for Atmospheric Research; James Hansen, National Aeronautics and Space Administration (NASA) Goddard Institute for Space Studies; Dennis Hartmann, University of Washington; Eugenia Kalnay, University of Maryland; Yoram Kaufman, NASA Goddard Space Flight Center; James Mahoney, U.S. Climate Change Science Program; Kenneth Mooney, NOAA; Richard Moss, U.S. Climate Change Science Program; V. Ramaswamy, NOAA Geophysical Fluid Dynamics Laboratory; Daniel Rosenfeld, Hebrew University; Susan Solomon, NOAA Aeronomy Laboratory; Graeme Stephens, Colorado State University; Lucia Tsaoussi, NASA; and Josh Willis, Scripps Institution of Oceanography.

The committee hopes that this report will be useful to the U.S. Climate Change Science Program in mapping future research directions to improve our knowledge of radiative and other climate forcings, their variability, and their impacts on climate.

Daniel J. Jacob
Chair

Acknowledgments

This report has been reviewed in draft form by individuals chosen for their diverse perspectives and technical expertise, in accordance with procedures approved by the National Research Council's Report Review Committee. The purpose of this independent review is to provide candid and critical comments that will assist the institution in making its published report as sound as possible and to ensure that the report meets institutional standards for objectivity, evidence, and responsiveness to the study charge. The review comments and draft manuscript remain confidential to protect the integrity of the deliberative process. We wish to thank the following individuals for their review of this report:

Tami C. Bond, University of Illinois
James A. Coakley, Jr., Oregon State University
Robert E. Dickinson, Georgia Institute of Technology
James A. Edmonds, Pacific Northwest National Laboratory
Jonathan A. Foley, University of Wisconsin
Peter R. Gent, National Center for Atmospheric Research
Richard Goody, Harvard University
Venkatachalam Ramaswamy, Geophysical Fluid Dynamics Laboratory
William J. Randel, National Center for Atmospheric Research
Stephen E. Schwartz, Brookhaven National Laboratory

Although the reviewers listed above have provided constructive comments and suggestions, they were not asked to endorse the report's conclusions or recommendations, nor did they see the final draft of the report before its release. The review of this report was overseen by John H. Seinfeld, California Institute of Technology. Appointed by the National Research Council, he was responsible for making certain that an independent examination of this report was carried out in accordance with institutional procedures and that all review comments were carefully considered. Responsibility for the final content of this report rests entirely with the authoring committee and the institution.

Contents

Executive Summary

The Earth receives a continuous influx of energy from the Sun. Some of this energy is absorbed at the Earth's surface or by the atmosphere, while some is reflected back to space. At the same time, the Earth and its atmosphere emit energy to space, resulting in an approximate balance between energy received and energy lost. Knowledge of the natural and anthropogenic processes that affect this energy balance is critical for understanding how Earth's climate has changed in the past and will change in the future.

In order to advance understanding of this issue, the U.S. Climate Change Science Program asked the National Academies to examine the current state of knowledge of how the energy balance regulating Earth's climate is modified by "forcings" including gases and aerosols, land use, and solar variability and to identify relevant research needs (see Appendix B for the full statement of task). This report provides the consensus view of the committee that was formed to undertake the study. In this report, the committee presents specific recommendations for expanding current radiative forcing concepts and metrics and outlines research priorities for exploiting these concepts and metrics as tools for climate change research and policy.

WHAT IS RADIATIVE FORCING?

Factors that drive climate change are usefully separated into forcings and feedbacks (Figure ES-1). A *climate forcing* is an energy imbalance imposed on the climate system either externally or by human activities.

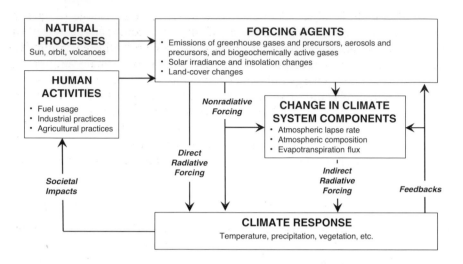

FIGURE ES-1 Conceptual framework of climate forcing, response, and feedbacks under present-day climate conditions. Examples of human activities, forcing agents, climate system components, and variables that can be involved in climate response are provided in the lists in each box.

Examples include changes in solar energy output, volcanic emissions, deliberate land modification, or anthropogenic emissions of greenhouse gases, aerosols, and their precursors. A *climate feedback* is an internal climate process that amplifies or dampens the climate response to a specific forcing. An example is the increase in atmospheric water vapor that is triggered by an initial warming due to rising carbon dioxide (CO_2) concentrations, which then acts to amplify the warming through the greenhouse properties of water vapor. Climate forcings are usefully subdivided into direct radiative forcings, indirect radiative forcings, and nonradiative forcings. *Direct radiative forcings* directly affect the radiative budget of the Earth; for example, added CO_2 absorbs and emits infrared (IR) radiation. *Indirect radiative forcings* create an energy imbalance by first altering climate system components (e.g., precipitation efficiency of clouds), which then lead to changes in radiative fluxes; an example is the effect of solar variability on stratospheric ozone. *Nonradiative forcings* create an energy imbalance that does not directly involve radiation; an example is the increasing evapotranspiration flux resulting from agricultural irrigation.

Studies of long-term changes in climate have emphasized global mean surface temperature as the primary index for climate change. The concept of "radiative forcing" provides a way to quantify and compare the contributions of different agents that affect surface temperature. Radiative forcing traditionally has been defined as the instantaneous change in energy flux at the tropopause resulting from a change in a component external to the climate system. Many current applications use an "adjusted" radiative forcing in which the stratosphere is allowed to relax to thermal steady state, thus focusing on the energy imbalance in the Earth and troposphere system, which is most relevant to surface temperature change. Once the stratosphere has been allowed to adjust to a forcing, the change in energy flux at the tropopause is equivalent to that at the top of the atmosphere (TOA), which is how radiative forcings are commonly reported.

Figure ES-2 shows the magnitude of several important forcings as esti-

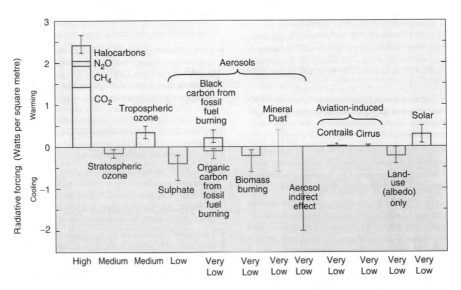

FIGURE ES-2 Estimated radiative forcing since preindustrial times for the Earth and troposphere system (TOA radiative forcing with adjusted stratospheric temperatures). The height of the rectangular bar denotes a central or best estimate of the forcing, while each vertical line is an estimate of the uncertainty range associated with the forcing, guided by the spread in the published record and physical understanding, and with no statistical connotation. Each forcing agent is associated with a level of scientific understanding, which is based on an assessment of the nature of assumptions involved, the uncertainties prevailing about the processes that govern the forcing, and the resulting confidence in the numerical values of the estimate. On the vertical axis, the direction of expected surface temperature change due to each radiative forcing is indicated by the labels "warming" and "cooling." SOURCE: IPCC (2001).

mated in the most recent synthesis report of the Intergovernmental Panel on Climate Change (IPCC, 2001). The largest positive forcing (warming) in Figure ES-2 is from the increase of well-mixed greenhouse gases (CO_2, nitrous oxide [N_2O], methane [CH_4], and chlorofluorocarbons [CFCs]) and amounted to 2.4 W m^{-2} (watts per square meter) between the years 1750 and 2000. Of the forcings shown in the figure, the radiative impact of aerosols is the greatest uncertainty.

The radiative forcing concept has been used extensively in the climate research literature over the past few decades and has also become a standard tool for policy analysis endorsed by the Intergovernmental Panel on Climate Change. For a wide range of forcings, there is a nearly linear relationship between the TOA radiative forcing and the resulting equilibrium response of global mean surface temperature as simulated in general circulation models. This allows quantitative and expedient comparison of the effects of different forcings in the past and of various possible future forcing scenarios. TOA radiative forcing is relatively easy to compute, generally robust across models, straightforward to use in policy applications, directly observable from space, and also inferable from observed changes in ocean heat content. It provides an extremely useful metric for climate change research and policy.

EXPANDING THE RADIATIVE FORCING CONCEPT

Despite all these advantages, the traditional global mean TOA radiative forcing concept has some important limitations, which have come increasingly to light over the past decade. The concept is inadequate for some forcing agents, such as absorbing aerosols and land-use changes, that may have regional climate impacts much greater than would be predicted from TOA radiative forcing. Also, it diagnoses only one measure of climate change—global mean surface temperature response—while offering little information on regional climate change or precipitation. These limitations can be addressed by expanding the radiative forcing concept and through the introduction of additional forcing metrics. In particular, the concept needs to be extended to account for (1) the vertical structure of radiative forcing, (2) regional variability in radiative forcing, and (3) nonradiative forcing. A new metric to account for the vertical structure of radiative forcing is recommended below. Understanding of regional and nonradiative forcings is too premature to recommend specific metrics at this time. Instead, the committee identifies specific research needs to improve quantification and understanding of these forcings.

Account for the Vertical Structure of Radiative Forcing

The relationship between TOA radiative forcing and surface temperature is affected by the vertical distribution of radiative forcing within the atmosphere. This effect is dramatic for absorbing aerosols such as black carbon, which may have little TOA forcing but greatly reduce solar radiation reaching the surface. It can also be important for land-use driven changes in the evapotranspiration flux at the surface, which change the energy deposited in the atmosphere without necessarily affecting the surface radiative flux. These effects can be addressed by considering surface as well as TOA radiative forcing as a metric of energy imbalance. The net radiative forcing of the atmosphere can be deduced from the difference between TOA and surface radiative forcing and may be able to provide information on expected changes in precipitation and vertical mixing. Adoption of surface radiative forcing as a new metric will require research to test the ability of climate models to reproduce the observed vertical distribution of forcing (e.g., from aircraft campaigns) and to investigate the response of climate to the vertical structure of the radiative forcing.

PRIORITY RECOMMENDATIONS:
  ❖ Test and improve the ability of climate models to reproduce the observed vertical structure of forcing for a variety of locations and forcing conditions.
  ❖ Undertake research to characterize the dependence of climate response on the vertical structure of radiative forcing.
  ❖ Report global mean radiative forcing at *both* the surface and the top of the atmosphere in climate change assessments.

Determine the Importance of Regional Variation in Radiative Forcing

Regional variations in radiative forcing may have important regional and global climatic implications that are not resolved by the concept of global mean radiative forcing. Tropospheric aerosols and landscape changes have particularly heterogeneous forcings. To date, there have been only limited studies of regional radiative forcing and response. Indeed, it is not clear how best to diagnose a regional forcing and response in the observational record; regional forcings can lead to global climate responses, while global forcings can be associated with regional climate responses. Regional diabatic heating can also cause atmospheric teleconnections that influence regional climate thousands of kilometers away from the point of forcing. Improving societally relevant projections of regional climate impacts will require a better understanding of the magnitudes of regional forcings and the associated climate responses.

PRIORITY RECOMMENDATIONS:

❖   Use climate records to investigate relationships between regional radiative forcing (e.g., land-use or aerosol changes) and climate response in the same region, other regions, and globally.

❖   Quantify and compare climate responses from regional radiative forcings in different climate models and on different timescales (e.g., seasonal, interannual), and report results in climate change assessments.

Determine the Importance of Nonradiative Forcings

Several types of forcings—most notably aerosols, land-use and land-cover change, and modifications to biogeochemistry—impact the climate system in nonradiative ways, in particular by modifying the hydrological cycle and vegetation dynamics. Aerosols exert a forcing on the hydrological cycle by modifying cloud condensation nuclei, ice nuclei, precipitation efficiency, and the ratio between solar direct and diffuse radiation received. Other nonradiative forcings modify the biological components of the climate system by changing the fluxes of trace gases and heat between vegetation, soils, and the atmosphere and by modifying the amount and types of vegetation. No metrics for quantifying such nonradiative forcings have been accepted. Nonradiative forcings have eventual radiative impacts, so one option would be to quantify these radiative impacts. However, this approach may not convey appropriately the impacts of nonradiative forcings on societally relevant climate variables such as precipitation or ecosystem function. Any new metrics must also be able to characterize the regional structure in nonradiative forcing and climate response.

PRIORITY RECOMMENDATIONS:

❖   Improve understanding and parameterizations of aerosol-cloud thermodynamic interactions and land-atmosphere interactions in climate models in order to quantify the impacts of these nonradiative forcings on both regional and global scales.

❖   Develop improved land-use and land-cover classifications at high resolution for the past and present, as well as scenarios for the future.

Provide Improved Guidance to the Policy Community

The radiative forcing concept is used extensively to inform climate policy discussions, in particular to compare the relative impacts of forcing agents. For example, integrated assessment models use radiative forcing as input to simple climate models, which are linked with socioeconomic models that predict economic damages from climate impacts and costs of various response strategies. The simplified climate models generally focus on global mean surface temperature, ignoring regional temperature changes and other societally relevant aspects of climate, such as rainfall or sea level. Incorporating these complexities is evidently needed in policy analysis. It is important to communicate the expanded forcing concepts as described in this report to the policy community and to develop the tools that will make their application useful in a policy context.

PRIORITY RECOMMENDATION:
* ❖ Encourage policy analysts and integrated assessment modelers to move beyond simple climate models based entirely on global mean TOA radiative forcing and incorporate new global and regional radiative and nonradiative forcing metrics as they become available.

ADDRESSING KEY UNCERTAINTIES

The radiative forcing since preindustrial times by well-mixed greenhouse gases is well understood. However, there are major gaps in understanding of the other forcings, as well as of the link between forcings and climate response. Error bars remain large for current estimates of radiative forcing by ozone, and are even larger for estimates of radiative forcing by aerosols. Nonradiative forcings are even less well understood. The following recommendations identify critical research avenues that should be persued immediately with high priority.

Conduct Accurate Long-Term Monitoring of Radiative Forcing Variables

The most important step for improving understanding of forcings is to obtain a robust record of radiative forcing variables, both in the past and into the future. A robust observational record is essential for improved understanding of the past and future evolution of climate forcings and responses. Existing observational evidence from surface-based networks, other in situ data (e.g., aircraft campaigns, ocean buoys), remote sensing platforms, and a range of proxy data sources (e.g., tree rings, ice cores) has enabled substantial progress in understanding, but there remain important shortcomings. The observational evidence needs to be more complete both in terms of the spatiotemporal and electromagnetic spectral coverage and in terms of the quantities measured. Long-term monitoring of forcing and other climate variables at much improved accuracy is needed to detect and understand future changes. In addition, surface-based observational networks for the detection of long term changes in climate variables need to be improved, notably by accounting for local changes (e.g., in land use and vegetation dynamics). Long-term, accurate observations of changes in the heat content of the oceans are also needed as a continuous record of globally averaged radiative forcing.

PRIORITY RECOMMENDATIONS:
   ❖   Continue observations of climate forcings and variables without interruption for the foreseeable future in a manner consistent with established climate monitoring principles (e.g., adequate cross-calibration of successive, overlapping datasets).
   ❖   Develop the capability to obtain benchmark measurements (i.e., with uncertainty significantly smaller than the change to be detected) of key parameters (e.g., sea level altimetry, solar irradiance, and spectrally resolved, absolute radiance to space).
   ❖   Conduct highly accurate measurements of global ocean heat content and its change over time.

Advance the Attribution of Decadal to Centennial Climate Change

Establishing relationships between past climate changes and known natural and anthropogenic forcings provides information on how such forcings may impact large-scale climate in the future. Instrumental records extend back about 150 years at best. Comparisons of observed surface temperatures with those simulated using reconstructions of the past forcings have yielded important insights into the roles of various natural and anthropogenic factors governing climate change. However, the shortness of the instrumental record limits the confidence with which climate change since preindustrial times can be attributed to specific forcings. Proxy records obtained from ice cores, sediments, tree rings, and other sources provide a critical tool for extending knowledge of forcings and effects further back in history. The lack of proxy climate data in certain key regions is a major limitation. Such regional information is important in evaluating the potential roles of changes in modes of climate variability, such as the El Niño/Southern Oscillation (ENSO).

PRIORITY RECOMMENDATIONS:
   ❖   Develop a best-estimate climate forcing history for the past century to millennium.
   ❖   Using an ensemble of climate models, simulate the regional and global climate response to the best-estimate forcings and compare to the observed climate record.

Reduce Uncertainties Associated with Indirect Aerosol Radiative Forcing

The interaction between aerosols and clouds can lead to a number of indirect radiative effects that arguably represent the greatest uncertainty in current radiative forcing assessments. In the so-called first indirect aerosol effect, the presence of aerosols leads to clouds with more but smaller particles; such clouds are more reflective and therefore have a negative radiative forcing. These smaller cloud droplets can also decrease the precipitation efficiency and prolong cloud lifetime; this is known as the second indirect aerosol effect. The so-called semidirect aerosol effect occurs when absorption of solar radiation by soot leads to an evaporation of cloud droplets. A number of research avenues hold promise for improving understanding of indirect and semidirect aerosol effects and for better constraining estimates of their magnitude. These include fundamental research on the physical and chemical composition of aerosols, aerosol activation, cloud microphysics, cloud dynamics, and subgrid-scale variability in relative humidity and vertical velocity.

PRIORITY RECOMMENDATION:
❖ Improve understanding and parameterizations of the indirect aerosol radiative and nonradiative effects in general circulation models using process models, laboratory measurements, field campaigns, and satellite measurements.

Better Quantify the Direct Radiative Effects of Aerosols

Aerosols have direct radiative effects in that they scatter and absorb both shortwave and longwave radiation. Knowledge of direct radiative forcing of aerosols is limited to a large extent by uncertainty about the global distributions and mixing states of aerosols. Mixing states have major implications on aerosol optical properties that are not well understood and are difficult to parameterize in climate models. Small-scale variability of humidity and temperature, which has a major impact on aerosol optical properties, is also difficult to represent in models. Mechanisms of aerosol production are not understood, so the effects of future changes in emissions and climate are highly uncertain. Removal of aerosols from the atmosphere occurs mainly by wet deposition, but model parameterizations of this process are highly uncertain and rudimentary in their coupling to the hydrological cycle.

PRIORITY RECOMMENDATIONS:
❖ Improve representation in global models of aerosol microphysics, growth, reactivity, and processes for their removal from the atmosphere through laboratory studies, field campaigns, and process models.
❖ Better characterize the sources and the physical, chemical, and optical properties of carbonaceous and dust aerosols.

Better Quantify Radiative Forcing by Ozone

Ozone is a major greenhouse gas. The greatest uncertainty in quantifying this forcing lies in reconstructing ozone concentrations in the past and projecting them into the future. Global modeling of tropospheric ozone remains a major challenge because of the complex coupling between photochemical and transport processes. The inability of models to reproduce ozone trends over the twentieth century suggests that there could be large errors in current estimates of natural ozone levels and the sensitivity of ozone to human influence. These errors could relate to emissions of precursors, chemical processes, and stratospheric influence. Lightning emissions of nitrogen oxides are particularly uncertain and play a major role in ozone production in the middle and upper troposphere where the radiative effect is maximum. Transport of ozone between the stratosphere and troposphere greatly affects upper tropospheric concentrations in a manner that is still poorly understood.

PRIORITY RECOMMENDATION:

❖   Improve understanding of the transport of ozone in the upper troposphere and lower stratosphere region and the ability of models to describe this transport.

Integrate Climate Forcing Criteria in Environmental Policy Analysis

Policies designed to manage air pollution and land use may be associated with unintended impacts on climate. Increasing evidence of health effects makes it likely that aerosols and ozone will be the targets of stricter regulations in the future. To date, control strategies have not considered the potential climatic implications of emissions reductions. Regulations targeting black carbon emissions or ozone precursors would have combined benefits for public health and climate. However, because some aerosols have a negative radiative forcing, reducing their concentrations could actually increase radiative warming. Policies associated with land management practices could also have inadvertent effects on climate. The continued conversion of landscapes by human activity, particularly in the humid tropics, has complex and possibly important consequences for regional and global climate change as a result of changes in the surface energy budget.

PRIORITY RECOMMENDATIONS:

❖   Apply climate models to the investigation of scenarios in which aerosols are significantly reduced over the next 10 to 20 years and for a range of cloud microphysics parameterizations.

❖   Integrate climate forcing criteria in the development of future policies for air pollution control and land management.

1

Introduction

T he climate of the Earth over its history has varied from "snowball" conditions with global ice cover to "hothouse" conditions when glaciers all but disappeared. Over the past 10,000 years (current interglacial, called the Holocene), the climate has been remarkably stable and favorable for human civilizations to flourish. Even during this stable period there have been notable regional climatic fluctuations such as the so-called Little Ice Age (A.D. 1600-1800), when Europe experienced unusually cold conditions. Increasing evidence points to a large human impact on global climate over the past decades through emissions of greenhouse gases and aerosols and through widespread changes in land cover (IPCC, 2001; NRC, 2001).

Climate change is driven by perturbations to the energy balance of the Earth system. These perturbations are called "climate forcings" and have been the subject of considerable scientific inquiry both for understanding Earth's history and for projecting future change. Indeed, further enhancing knowledge of climate forcings is critical for improving projections of future climate change. It is in that context that the U.S. Climate Change Science Program asked the National Academies to examine the current state of knowledge regarding the climate forcings associated with gases, aerosols, land use, and solar variability and to identify relevant research needs (see Appendix B for full statement of task). In response, the Committee on Radiative Forcing Effects on Climate was formed (see Appendix A for biographies of committee members). This report provides the committee's consensus views on the current understanding of different climate forcings, considers alternatives for quantifying and comparing different forcing agents, and recommends research priorities for attaining a more complete understanding of climate forcing.

EARTH'S CLIMATE SYSTEM

Climate is conventionally defined as the long-term statistics of the weather (e.g., temperature, cloudiness, precipitation). This definition emphasizes the atmospheric and physical components of the climate system. These physical processes within the atmosphere are affected by ocean circulation, the reflectivity of the Earth's surface, the chemical composition of the atmosphere, and vegetation patterns, among other factors. Improved understanding of how the atmosphere interacts with the oceans, the cryosphere (ice-covered regions of the world), and the terrestrial and marine biospheres has led scientists to expand the definition of climate to encompass the oceanic and terrestrial spheres as well as chemical components of the atmosphere (Figure 1-1). This expanded definition promotes an Earth system approach to studying how and why climate changes.

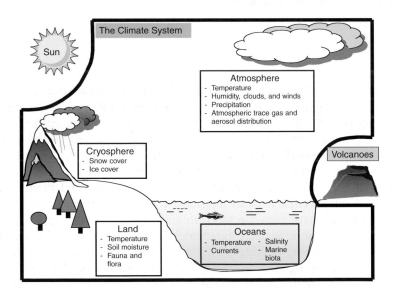

FIGURE 1-1 The climate system, consisting of the atmosphere, oceans, land, and cryosphere. Important state variables for each sphere of the climate system are listed in the boxes. For the purposes of this report, the Sun, volcanic emissions, and human-caused emissions of greenhouse gases and changes to the land surface are considered external to the climate system.

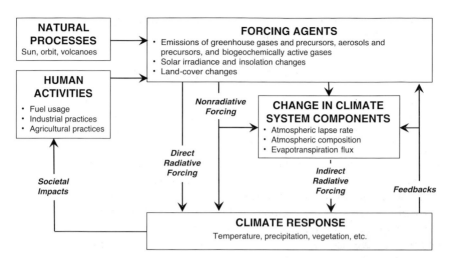

FIGURE 1-2 Conceptual framework of climate forcing, response, and feedbacks under present-day climate conditions. Examples of human activities, forcing agents, climate system components, and variables that can be involved in climate response are provided in the lists in each box.

Factors that affect climate change are usefully separated into forcings and feedbacks. The conceptual diagram of Figure 1-2 illustrates the connections between climate forcings, responses, and feedbacks as defined in this report. A *climate forcing* is an energy imbalance imposed on the climate system either externally or by human activities. Examples include changes in solar energy output, volcanic emissions, deliberate land modification, or anthropogenic emissions of greenhouse gases, aerosols, and their precursors. A *climate feedback* is an internal climate process that amplifies or dampens the climate response to an initial forcing. An example is the increase in atmospheric water vapor that is triggered by an initial warming due to rising carbon dioxide (CO_2) concentrations, which then acts to amplify the warming through the greenhouse properties of water vapor. Climate change feedbacks are the subject of a recent report of the National Research Council (NRC, 2003).

Climate forcings can be classified as radiative (direct or indirect) or nonradiative. *Direct radiative forcings* affect the radiative budget of the Earth directly; for example, added CO_2 absorbs and emits infrared (IR) radiation. *Indirect radiative forcings* create a radiative imbalance by first altering climate system components, which then almost immediately lead to changes in radiative fluxes; an example is the effect of aerosols on the precipitation efficiency of clouds. *Nonradiative forcings* create an energy imbalance that does not involve radiation directly; an example is the increasing evapotranspiration flux resulting from agricultural irrigation. This report focuses on the forcing agents and the ways in which they act to create a climate response (i.e., downward arrows in Figure 1-2). Although they are not the primary focus of this report, it is necessary at times to address climate responses because of what they tell us about climate forcings. See Box 1-1 and Appendix C for definitions of important terms.

Some times the climate system is defined more broadly by including the Sun, the lithosphere (the Earth's crust), or even humans as part of the climate system (e.g., Claussen, 2004; Steffen et al., 2004). For the purposes of this report, however, those elements that impact climate but are not affected by it are considered external to the climate system. Changes in solar output are viewed as a natural external forcing because the Earth does not affect the Sun. Volcanic aerosols are also considered a natural external forcing because Earth's climate does not impact volcanic activity, except on very long timescales. Increases in CO_2 and other greenhouse gases due to human activities are assumed to be an external anthropogenic forcing. Defining the climate system in this way allows separation between external climate forcings and internal climate responses to those forcings. This definition is consistent with that adopted in the recent NRC report on climate change feedbacks (NRC, 2003).

The definition of climate forcing and climate response may vary depending on the timescale under consideration. On the timescale of billions of years, greenhouse gas concentrations may both influence climate, through their radiative properties, and be influenced by climatic variations in weathering rates. On the timescale of millions of years, on the other hand, greenhouse gas concentrations are determined largely by slowly evolving tectonic boundary conditions. In this case, greenhouse gas concentrations can be treated as a forcing, and changes in global mean temperature can be considered a response. Over the past 1000 years, CO_2 concentrations appear to have varied in response to surface temperature changes prior to large-scale fossil fuel burning during the nineteenth and twentieth centuries, while during the latter period they can be considered primarily as a forcing of surface temperature changes (Gerber et al., 2003).

Given the conventional focus of climatologists on temperature as well as the clear link between greenhouse gases and surface temperature, studies of long-term changes in climate have emphasized temperature as the primary index for climate change. The concept of "radiative forcing" provides a way to quantify and compare the contributions of different agents that affect surface temperature by modifying the balance between incoming and

BOX 1-1
Key Definitions

Climate system: The system consisting of the atmosphere, hydrosphere, lithosphere, and biosphere, determining the Earth's climate as the result of mutual interactions and responses to external influences (forcing). Physical, chemical, and biological processes are involved in the interactions among the components of the climate system.

Climate forcing: An energy imbalance imposed on the climate system either externally or by human activities.

- *Direct radiative forcing:* A climate forcing that directly affects the radiative budget of the Earth's climate system; for example, added carbon dioxide (CO_2) absorbs and emits infrared radiation. Direct radiative forcing may be due to a change in concentration of radiatively active gases, a change in solar radiation reaching the Earth, or changes in surface albedo. Radiative forcing is reported in the climate change scientific literature as a change in energy flux at the tropopause, calculated in units of watts per square meter (W m^{-2}); model calculations typically report values in which the stratosphere was allowed to adjust thermally to the forcing under an assumption of fixed stratospheric dynamics.
- *Indirect radiative forcing:* A climate forcing that creates a radiative imbalance by first altering climate system components (e.g., precipitation efficiency of clouds), which then almost immediately lead to changes in radiative fluxes. Examples include the effect of solar variability on stratospheric ozone and the modification of cloud properties by aerosols.
- *Nonradiative forcing:* A climate forcing that creates an energy imbalance that does not immediately involve radiation. An example is the increasing evapotranspiration flux resulting from agricultural irrigation.

Climate response: Change in the climate system resulting from a climate forcing.

Climate feedback: An amplification or dampening of the climate response to a specific forcing due to changes in the atmosphere, oceans, land, or continental glaciers.

NOTE: Additional definitions are provided in Appendix C.

outgoing radiative energy fluxes. Global radiative forcing at the top of the atmosphere (TOA), as used in assessments by the Intergovernmental Panel on Climate Change (IPCC), is relatively easy to compute in climate models and has straightforward policy applications. However, it has important limitations when applied to radiative forcing agents not conventionally considered as such (e.g., aerosols, land-use change) or when used to measure climatic implications other than global mean temperature (e.g., regional precipitation). To address these limitations, the concept of radiative forcing needs to be expanded; that expansion is a major theme of this report.

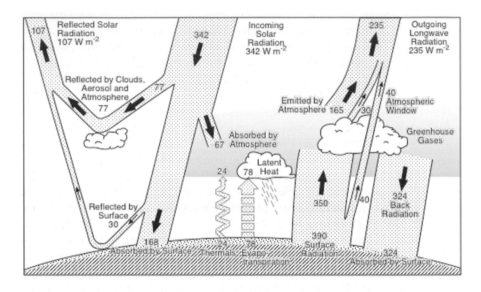

FIGURE 1-3 Energy budget for the atmospheric components of the climate system. SOURCE: Kiehl and Trenberth (1997).

RADIATIVE FORCING:
PERTURBATION TO EARTH'S ENERGY BUDGET

The various physical processes that contribute to Earth's global annual mean energy budget are shown in Figure 1-3. The Earth receives a continuous influx of energy from the Sun. About 69 percent of this energy is absorbed at the Earth's surface or by the atmosphere, while the rest is reflected back to space. At the same time, the Earth and its atmosphere emit energy to space, resulting in an approximate balance between energy received and energy lost. The so-called greenhouse gases, such as water vapor, CO_2, methane (CH_4), nitrous oxide (N_2O), halocarbons, and ozone (O_3), modify this balance by absorbing and then re-emitting some of the outgoing radiation.[1] Small particles in the atmosphere (aerosols) also ab-

---

[1]Each object in the Earth system (e.g., a gas molecule in the atmosphere, the surface of the Earth itself) absorbs and emits energy at a rate that depends on the object's temperature. For this reason, greenhouse gases are most effective when located in parts of the atmosphere that are much colder than the Earth's surface, such as the upper troposphere. Greenhouse gases present in this region absorb energy emitted by the surface, but emit less energy to space because they are much colder than the surface.

sorb and scatter some of the outgoing radiation. Changes in the Sun or in the reflective characteristics of the Earth's surface represent additional perturbations. Knowledge of the natural and anthropogenic processes that affect Earth's energy balance is critical for understanding how Earth's climate has changed in the past and will change in the future

The energy budget in Figure 1-3 assumes an exact balance in energy fluxes. In reality, the solar flux, the chemical composition of Earth's atmosphere, and surface conditions maintain a perpetual state of slight imbalance. For example, solar flux is affected by variability in Earth's orbit around the Sun and by changes in the intensity of solar output. Radiative forcing provides a conceptual framework for thinking about how Earth's energy budget can be modified and for quantifying the modifications and their potential impact in terms of surface temperature response. Radiative forcing has traditionally been defined as a change in energy flux at the tropopause resulting from a change in a component external to the physical climate system (see Box 1-2). This definition was used initially in radiative transfer models, in which its application is straightforward (Manabe and Strickler, 1964; Manabe and Wetherald, 1967). In these models, the internal climate state (e.g., thermal structure, clouds, water vapor) is held fixed and the only change made to the model is in a single atmospheric constituent (e.g., CO_2 mixing ratio). This constitutes a change in radiative flux with all internal variables held fixed or, in mathematical terms, the partial derivative of the radiative flux with respect to a single constituent.

Initial applications of radiative forcing focused on the immediate change in the radiative flux at the top of the atmosphere. It was soon realized that the stratosphere, the layer of the atmosphere extending from about 10 to 50 km altitude, returns rapidly (in about a year) to radiative equilibrium and is largely decoupled from the surface. In contrast, the surface and the troposphere (the lowest layer of the atmosphere) are strongly coupled through vertical motions. The definition of TOA radiative forcing was thus modified to allow the stratosphere to return to local thermal equilibrium. Such "adjusted radiative forcing" values are the standard reported in the literature today. They are sometimes referred to as the radiative forcing at the tropopause (boundary between the troposphere and the stratosphere), although with a stratosphere in radiative equilibrium this is equivalent to TOA radiative forcing. Stratospheric adjustment is most important for forcings that affect the stratospheric thermal structure, such as well-mixed greenhouse gases. For nonuniform perturbations near the tropopause the adjustment can be quite sensitive to the vertical profile of forcing.

The original estimation of radiative forcing was carried out in terms of a change in the globally averaged radiative flux. For well-mixed gases this provides a reasonable estimate of forcing, but even for these types of gases there is still considerable latitudinal variation in radiative forcing from pole

---

BOX 1-2
Bucket Analogy for Radiative Forcing

A steady state climate system exists when the amount of energy entering the system equals the amount of energy leaving the system. For Earth this means that the amount of solar energy absorbed by the surface-troposphere system equals the amount of longwave energy emitted to space. A simple analogy is that of a bucket with a hole in it. Imagine water flowing into the bucket at some fixed rate. The water will flow out of the bucket at a rate dependent on the size of the hole in the bucket and the depth of the water. The water in the bucket will reach a fixed level once the amount of water leaving the bucket equals the amount entering the bucket. The water flowing into the bucket is analogous to the solar energy flowing into Earth's system. The amount of water flowing out of the bucket is analogous to the longwave energy leaving Earth's system. The level of water in the bucket is analogous to Earth's global heat content (which is determined to a first order by mean temperature and water vapor pressure).

The concept of radiative forcing is analogous to a change in either the amount of water flowing into the bucket or the amount of water leaving the bucket. For example, if the hole in the bucket was made smaller, less water would flow out initially and the water level would rise until there was a new balance between what flows into and out of the bucket. This is analogous to increasing a greenhouse gas in Earth's atmosphere, impeding the escape of longwave radiation to space. An increase in the flow of water into the bucket will also result in an increase in the water level, which is analogous to an increase in solar energy. Both changes lead to increased global heat content, all other factors remaining constant. Radiative forcing is a change in the amount of energy per unit time flowing into or out of Earth's climate system.

---

to equator, which is due mainly to variation in temperature and specific humidity (e.g., Kiehl and Ramanathan, 1982). For tropospheric ozone, tropospheric aerosols, and land surface changes, there is considerable spatial variability in radiative forcing (e.g., Kiehl and Briegleb, 1993; Kiehl et al., 1998; Pielke et al., 2002). Some have argued that spatial variation in forcing is dampened in the climate response because of the homogenization effect from atmospheric transport of heat. However, it is becoming apparent that spatial variations in forcing are important to understanding observed climate signals (e.g., Karl et al., 1995; Hegerl et al., 2003).

Until five years ago the focus in radiative forcing was on changes in tropopause or TOA fluxes. However, early studies on absorbing aerosols (e.g., Ogren and Charlson, 1983) recognized that these aerosols would require a focus on radiative forcing at the Earth's surface. Recent observations and consequent modeling studies from the Indian Ocean Experiment (INDOEX) found that the presence of absorbing aerosols led to significantly different changes in shortwave radiative flux between the surface and the tropopause because a significant amount of energy is absorbed within the atmosphere (Ramanathan et al., 2001a). As shown in Chapter 4 (see Box 4-1) the TOA forcing observed during INDOEX over the Indian subcontinent was close to zero, but the surface forcing was -14 W m^{-2} and tropospheric forcing was about $+14$ W m^{-2}. As shown by several general circulation model (GCM) sensitivity studies (Ramanathan et al., 2001b; Chung and Ramanathan, 2003; Menon et al., 2002b), in spite of the near-zero TOA forcing, the introduction of absorbing aerosols results in a large surface cooling (-0.5 to -1 K) of the North Indian Ocean and South Asia, a large lower tropospheric warming (0.5 to 1 K), and large changes in regional precipitation. This means that in addition to calculating radiative forcing at the tropopause, one must also quantify radiative forcing at the surface and its atmospheric distribution.

THEORETICAL DEVELOPMENT OF THE RADIATIVE FORCING CONCEPT

The concept of radiative forcing is based on the hypothesis that the change in global annual mean surface temperature is proportional to the imposed global annual mean forcing, independent of the nature of the applied forcing. The fundamental assumption underlying the radiative forcing concept is that the surface and the troposphere are strongly coupled by convective heat transfer processes; that is, the earth-troposphere system is in a state of radiative-convective equilibrium (RCE; see Box 1-3). In the present context, the term "convective heat transfer" refers to heat transport by all types of vertical motions ranging from small (few meters) to planetary scales. The net result of radiative-convective equilibrium is that the vertical temperature profile within the troposphere (the so-called lapse rate) is largely determined by convective heat transport, while the vertically averaged surface-troposphere temperature is regulated by radiative flux equilibrium at the tropopause. RCE models were initially used to determine the vertical temperature profile of stellar (Chandrasekhar, 1947; Ambartsumyan, 1958) and planetary atmospheres (e.g., Chamberlain, 1960; Gierasch and Goody, 1968; Cess, 1972). Its first application to the Earth's atmosphere with a proper treatment of convective heat transport and the radiative transfer effects of infrared active gases and clouds was published by Manabe and Strickler (1964) and Manabe and Wetherald (1967).

According to the radiative-convective equilibrium concept, the equation for determining global average surface temperature of the planet is

$$\frac{dH}{dt} = f - \frac{T'}{\lambda} \qquad (1\text{-}1)$$

BOX 1-3
Radiative-Convective Interactions Between the
Surface and the Atmosphere

According to the RCE concept, radiative processes cool the troposphere and warm the ground. As shown in the figure below, the primary source of tropospheric cooling is infrared emission (or radiative cooling) by water vapor and clouds, while the ground warming is due to solar heating and back radiation from atmospheric water vapor and clouds. Such a pattern of atmospheric cooling and surface warming leads to superadiabatic lapse rates (temperature decreasing by more than 9.8 K km^{-1}) and triggers atmospheric convection. The ensuing vertical motions transport heat from the surface to the atmosphere and restore the lapse rate to neutral (adiabatic). The heat is released in the form of latent heating during condensation or as sensible heat from turbulent eddies originating in the boundary layer.

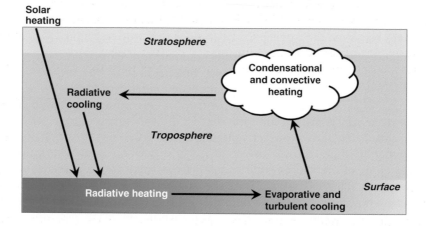

Schematic illustration of the balance between radiative and nonradiative processes under RCE. Arrows indicate direction of energy transfer. Within the troposphere, longwave radiative cooling far exceeds solar heating, resulting in a net radiative cooling. This cooling is balanced by release of latent heating by condensation and precipitation and convective transport of sensible heat transfer from the surface. At the surface, solar heating far exceeds longwave cooling, and this radiative heating is balanced by convective transport of latent and sensible heat from the surface to the atmosphere. In addition, the sum of surface radiative heating and tropospheric radiative cooling is zero, thus maintaining radiation energy balance for the whole surface-troposphere column, and it is this radiation balance that is perturbed by the addition of greenhouse gases and aerosols. Likewise, the sum of convective cooling of the surface and convective heating of the troposphere balance each other, and this balance is perturbed by land surface changes. SOURCE: Adapted from Ramanathan et al. (1989).

where

$$H = \int_{z_b}^{\infty} \rho C_p T dz \qquad (1\text{-}2)$$

is the heat content of the land-ocean-atmosphere system with ρ the density, C_p the specific heat, T the temperature, and z_b the depth to which the heating penetrates. Equation 1-1 describes the change in the heat content where f is the radiative forcing at the tropopause, T' is the change in surface temperature in response to a change in heat content, and λ is the climate feedback parameter (Schneider and Dickinson, 1974), also known as the climate sensitivity parameter, which denotes the rate at which the climate system returns the added forcing to space as infrared radiation or as reflected solar radiation (by changes in clouds, ice and snow, etc.). In essence, λ accounts for how feedbacks modify the surface temperature response to the forcing. In principle, T' should account for changes in the temperature of the surface and the troposphere, and since the lapse rate is assumed to be known or is assumed to be a function of surface temperature, T' can be approximated by the surface temperature. For steady state, the solution yields

$$T' = f\lambda \qquad (1\text{-}3)$$

Studies from one-dimensional radiative convective models initially indicated that λ was a nearly invariant parameter for a variety of forcings (Ramanathan et al., 1985). This finding has generally been supported by three-dimensional modeling studies of climate sensitivity, which indicate that λ varies by only about 25 percent within a particular model, although there can be much greater differences between models (Manabe and Wetherald, 1975; Hansen et al., 1997; IPCC, 2001).

The implication of Equation 1-3 with fixed λ is that surface temperature change is uniquely determined by the radiative forcing at the tropopause (or at the top of the atmosphere if the stratosphere is adjusted for radiative equilibrium). The primary validity for this concept was provided by a series of sensitivity studies by Manabe and Wetherald (1975), who used a three-dimensional GCM to calculate the global mean surface temperature change due to a doubling of CO_2 and a 2 percent change in solar insolation. They found that surface temperature estimated by the GCM can be scaled with the initial forcing as in Equation 1-3. Since then, similar sensitivity studies have been performed by other GCMs and basically confirmed Manabe and Wetherald's result, with most success for perturbations due to uniformly mixed greenhouse gases, changes in incoming solar irradiance, and homogeneously distributed scattering aerosols in the troposphere and the stratosphere (e.g., Cess and Potter, 1988; Hansen et al., 1997).

Highly inhomogeneous perturbations do not show as good a scaling. In addition, successful scaling for global mean surface temperature does not necessarily imply similar success for other climate variables such as precipitation (e.g., Chen and Ramaswamy, 1996).

The RCE concept and the formulation of Equation 1-3 for radiative forcing and feedback have thus been used since the 1960s, implicitly (Budyko, 1969; Kellogg and Schneider, 1974) as well as explicitly (Moller, 1963; Yamamoto and Tanaka, 1972; Schneider and Mass, 1975; Cess, 1976; and hundreds of studies since then). The term radiative forcing was not in vogue until the 1980s, although the climate forcing, feedback, and response framework was implicit in the 1979 "Charney Report," the NRC's first report addressing the potential for human-caused climate change (NRC, 1979). Early references introducing this terminology for anthropogenic changes are seen in Ramanathan et al. (1985) and the World Meteorological Organization report on atmospheric ozone (WMO, 1985; republished as Ramanathan et al., 1987), which also gives a detailed and still-accepted definition of radiative forcing. Early use of the term radiative forcing also can be found in the NRC report *Toward an Understanding of Global Change* (NRC, 1988) and the IPCC reports published in 1990, 1992, and 1996 (IPCC, 1990, 1992, 1996). During the early phases of its use, radiative forcing was not restricted to changes external to the climate system but was used in a more general sense. For example, the radiative effects of clouds were referred to as "cloud-radiative forcing" (Charlock and Ramanathan, 1985) because clouds introduce spatial and temporal gradients in radiative heating. After the IPCC Second Assessment Report (IPCC, 1996), the term radiative forcing came to imply climate forcing (i.e., a term that *forces* climate changes).

An important construct in the formulation of the concept of forcing is that the equilibrium of the Earth system must have matching incoming and outgoing energy fluxes. Over long time periods the proposition that the Earth is in thermal equilibrium with its surroundings must hold, but the timescale for relaxation of the entire system to this equilibrium may be as long as 2000-3000 years[2] for some forcings and the associated feedbacks. Current understanding of the climate response to radiative forcing relies heavily on climate equilibrium simulations rather than on transient responses. The uncertainty and variability of the predicted short-term (i.e., less than a couple of years) climate changes are high since many of the important processes parameterized in climate models have been tuned to

---

[2]A relaxation timescale of 2000-3000 years is determined by the deep ocean mixing timescale. If only the upper ocean is considered, the system can reach equilibrium in about 50 years, which corresponds to an *e*-folding response time of 2-3 years.

match climate statistics over decadal timescales. Simulating short-term regional climate change requires methods of downscaling large-scale information. Currently this is accomplished either through statistical approaches or by forcing a regional model with boundary conditions from a global climate model.

THE RADIATIVE FORCING CONCEPT AND CLIMATE POLICY

The concept of radiative forcing has provided a clear mechanism for conceptualizing the Earth's climate as a closed system with a detectable metric of change: global mean surface temperature. The metric is easily understandable and readily correlated to global-scale geological, oceanographic, and biological changes (e.g., ice caps melting, sea level rising, ecosystems changing). The projected changes in climate have been translated into economic costs (with varying uncertainty), providing a direct relationship between radiative forcing and economic impacts. This relationship enables policy makers to consider the relative benefits of investments in new technologies, emissions regulations, carbon taxes, sequestration and offsetting, and emissions trading. This conceptual framework is illustrated in Figure 1-4.

For most policy applications, the relationship between radiative forcing and surface temperature is assumed to be linear, thereby making it possible to add different forcings to assess the overall climate impact. As discussed above, the linearity of response in several GCM experiments using the radiative forcing of homogeneously distributed greenhouse gases supports this approach as do summary diagrams compiled to compare different radiative forcings, such as those presented in the IPCC reports (see Figure 2-1 of this report). In these diagrams, it is often assumed that the bars from different sources may be added to give an overall effect although this is not entirely correct.

The simplification of complex, mechanistically disparate processes to the same radiative forcing metric, with the implication that positive forcings may cancel negative forcings, provides a way of easily communicating climate forcing factors and their relative importance to general audiences. However, a net zero global mean radiative forcing may be associated with large regional or nonradiative (e.g., precipitation) changes. Further, when forcings are added, uncertainties in individual forcings must be propagated, resulting in large uncertainties in the total forcing. Adding forcings also belies the complexity of the underlying chemistry, physics, and biology. It suggests that all effects on climate can be quantified by a similar metric without knowing, or needing to know, the details of the climate response as captured in feedback effects. Yet there are many aspects of climate change—including rainfall, biodiversity, and sea level—that are currently not related quantitatively, much less linearly, to radiative forcings.

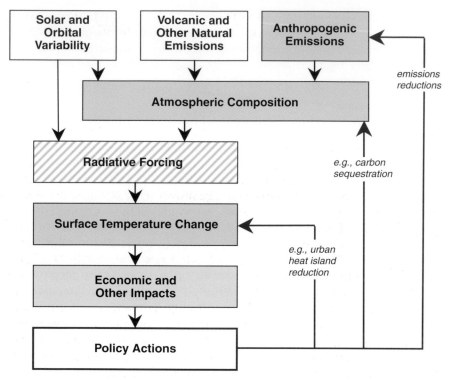

FIGURE 1-4 Conceptual framework for how radiative forcing fits into the climate policy framework. Blue-shaded boxes indicate quantities that have been considered as policy targets in international negotiations and other policy analyses. Radiative forcing (striped box) has not been treated as a policy target in the same explicit way that limiting emissions (e.g., Kyoto Protocol), limiting concentrations (e.g., greenhouse gas stabilization scenarios), and limiting temperature changes and impacts (e.g., environmental scenarios) have. That is, an explicit cap on anthropogenic radiative forcing levels has not been proposed analogous, for example, to the Kyoto Protocol cap on emissions. Note that land-use change has not received much attention as a forcing agent and is not included here, though this report recommends that it should be.

Simple Climate Models

The radiative forcing concept has been employed in simple climate models that rely on the assumption that climate sensitivity is constant. These models often use formulas for the radiative forcing for individual greenhouse gases, such as those published by the IPCC. A well-known example is the formula for the forcing $f(t)$ for CO_2 expressed in units of watts per square meter with a coefficient from IPCC (2001):

$$f(t) = 5.35 \ln \left[\frac{CO_2(t)}{CO_2(1750)} \right] \tag{1-4}$$

where $CO_2(t)$ is the atmospheric concentration of CO_2 for year t. Such models can relate greenhouse gas emissions to the equilibrium global averaged temperature changes and, using transient oceanic heat uptake models, to transient temperature changes and impacts. For short-lived species, such as aerosols, expressions of the type of Equation 1-4 are not available due to the great spatial variability in concentrations and optical properties.

When linked to socioeconomic models, simple climate models have become a powerful tool for policy analysis, often referred to as "integrated assessment" models (Manne et al., 1995; de Vries et al., 2000; Nordhaus and Boyer, 2000; Roehrl and Riahi, 2000; Matsuoka et al., 2001). They can be linked to an economic "damage" function that simulates the economic impacts and damages of global warming. The system may be further coupled to an optimization scheme to determine optimal investment rates in reductions of greenhouse gas emissions. In these optimization schemes the damage function is dependent on the global averaged surface temperature, while the cost function depends on the level of greenhouse gas emissions abatement (Nordhaus and Boyer, 2000). The simplified "box" approach to climate modeling used in most integrated assessment models is subject to criticism for ignoring regional temperature changes. The current dearth of regionally specific data on damages and their economic costs is a key limitation as well.

Integrated assessment models have been used to evaluate many climate policy questions. Most recently they have been part of the burgeoning suite of studies that evaluated the Kyoto Protocol treaty for greenhouse gas emissions targets for the years 2010-2015 (Kyoto Protocol, 1997). Various emissions pathways were studied in terms of their overall cost-benefit ratios (Nordhaus and Boyer, 2000). The policy targets can be limits on emissions rates as in the Kyoto Protocol; limits on greenhouse gas concentrations, which is the approach for stabilization studies; or limits on the rate of global warming as in the case of environmentally oriented scenarios. These possible policy targets are shown by the shaded boxes in Figure 1-4. Although radiative forcing is inherent to integrated assessment models, radiative forcing per se has not been treated as a climate policy target.

Global Warming Potentials and Greenhouse Gas Equivalence Models

Many climate policy questions require comparing the climate change effects of different greenhouse gases, aerosols, and other forcings. Such comparisons are integral to the formulation of climate treaties and the assessment of progress toward greenhouse gas emissions reductions. For example, if one party to a climate target achieves emissions reductions in CO_2, and another party focuses on CH_4, some metric is needed to compare these reductions in order to assess overall progress toward the target.

Policy analysts have sought a simple basis for quantitatively comparing the radiative consequences of emissions of different gases. The concept of global warming potential (GWP) was developed to address this need. GWPs compare the integrated radiative impact of a one time-unit of emissions of greenhouse gas X to the integrated radiative forcing impact of a one time-unit of CO_2 emissions (IPCC, 2001). Mathematically, GWP is expressed as

$$GWP(X) = \frac{\int_0^{TH} a_x \cdot [X(t)]dt}{\int_0^{TH} a_{CO_2} \cdot [CO_2(t)]dt} \qquad (1\text{-}5)$$

where TH is the time horizon over which the calculation is considered and a_x is the radiative efficiency of gas X, or the increase in radiative forcing for a unit increase in the atmospheric abundance of the substance. This radiative efficiency is typically expressed in units of W m^{-2} kg^{-1}. The parameter $X(t)$ is the time decay profile for the gas following its release into the atmosphere. The corresponding factors for CO_2, the reference gas, are in the denominator. Scaling the radiative impact of other forcings by that of CO_2 makes it easier to compare forcings quantitatively to each other, but this approach has been criticized because it depends on how well the radiative impact of CO_2 is understood. A change in the denominator of Equation 1-5 requires that the whole set of GWPs be revised, potentially introducing confusion.

The radiative forcing formulas (e.g., Equation 1-4) are used in the calculation of the efficiency term a_x. The marginal increase in radiative forcing can be calculated as the first derivative of the radiative forcing with respect to concentration. For low-concentration gases, such as chlorofluorocarbons (CFCs), whose radiative forcing increases linearly with concentration, this derivative is a constant. For more abundant gases such as CO_2, the derivative—and marginal radiative forcing response—depends on the background atmospheric concentration at the time of the hypothetical pulse release of the gas. The IPCC calculates GWPs for the well-mixed gases for three discrete time horizons of 20, 100, and 500 years. The Kyoto Protocol recommends that parties to the treaty use the 100-year value for comparing emissions reductions of different gases toward meeting targeted greenhouse emissions reductions for the first commitment period of 2008-2012 (Kyoto Protocol, 1997).

Application of the GWP concept has mainly been restricted to the long-lived greenhouse gases. In principle, it could be applied to short-lived forcing agents such as ozone and aerosols or, more specifically, to the emissions of their precursors (e.g., Schwartz, 1993), but there are a number of complicating factors including (1) the often poorly defined relationship between the precursor and the radiative forcing agent; (2) the inhomogeneity of the forcing; and (3) the much shorter time horizons (decades or less) relevant to the radiative forcing from these short-lived agents. In addition, the current concept is not useful for evaluating how the rate of technical transformation, which depends on economic and policy drivers, affects the trade-off between two greenhouse gases. At present, integrated assessment models are used to consider the combined scientific and economic factors that contribute to the global warming impacts of different forcings (e.g., Manne and Richels, 2001).

Many criticisms of the oversimplicity of the GWP approach have been published (Lashof, 2000; O'Neill, 2000; Smith and Wigley, 2000a,b). More complex equivalence calculations, such as the "forcing equivalence index" of Wigley (1998), have been developed to address its shortcomings. The essence of the forcing equivalence index is that a time series of emissions of a greenhouse gas produces a time series of radiative forcings. By inverting this temporal profile of radiative forcing in terms of the atmospheric properties of another greenhouse gas, the "equivalent" emissions of the alternative gas are estimated. This calculation, like the GWP, does not fully treat the complexities of the long-term behavior of the two gases (O'Neill, 2000).

2

State of Scientific Understanding

In this chapter the state of understanding of radiative forcing from individual agents is reviewed. Over the past 15 years, the Intergovernmental Panel on Climate Change (IPCC) has produced assessments in which at least one chapter has been devoted to a thorough review of current understanding about radiative forcings. The discussions here summarize the findings of the IPCC's Third Assessment Report (IPCC, 2001) and scientific advances since it was published.

The Third Assessment Report (IPCC, 2001) includes a summary figure of the global and annual mean radiative forcings from 1750 to 2000 due to a range of perturbations (Figure 2-1), including the well-mixed greenhouse gases, ozone, aerosols, aviation effects on clouds, land use, and the Sun. The largest positive forcing (warming) since 1750 is associated with the increase of the well-mixed greenhouse gases (carbon dioxide [CO_2]; nitrous oxide [N_2O]; methane [CH_4]; and chlorofluorocarbons [CFCs]) and amounts to 2.4 W m^{-2}. The greatest uncertainty in Figure 2-1 is associated with the direct and indirect radiative effects of aerosols. If the actual negative forcing from aerosols were at the high end (most negative) of the uncertainty range, then it would have offset essentially all of the positive forcing due to greenhouse gases (see also Boucher and Haywood, 2001).

According to the IPCC definition, applied to the data in Figure 2-1, "The radiative forcing of the surface-troposphere system due to the perturbation in or the introduction of an agent is the change in net irradiance at the tropopause after allowing for stratospheric temperatures to readjust to radiative equilibrium, but with the surface and tropospheric temperatures and state held fixed at the unperturbed values." This definition of forcing is

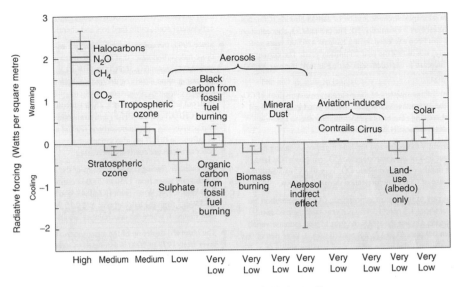

FIGURE 2-1 Estimated radiative forcings since preindustrial times for the Earth and troposphere system (TOA radiative forcing with adjusted stratospheric temperatures). The height of the rectangular bar denotes a central or best estimate of the forcing, while each vertical line is an estimate of the uncertainty range associated with the forcing, guided by the spread in the published record and physical understanding, and with no statistical connotation. Each forcing agent is associated with a level of scientific understanding, which is based on an assessment of the nature of assumptions involved, the uncertainties prevailing about the processes that govern the forcing, and the resulting confidence in the numerical values of the estimate. On the vertical axis, the direction of expected surface temperature change due to each radiative forcing is indicated by the labels "warming" and "cooling." SOURCE: IPCC (2001).

restricted to changes in the radiation balance of the Earth-troposphere system imposed by external factors, with no changes in stratospheric dynamics, without any surface and tropospheric feedbacks in operation, and with no dynamically induced changes in the amount and distribution of atmospheric water. A somewhat broader perspective is applied in this chapter to include, in particular, volcanic aerosols, the effects of land-use changes and aerosols on precipitation, and the radiative forcing due to changes in ocean color.

Figure 2-1 has been an effective way to portray the relative magnitudes of different radiative forcings, the associated scientific uncertainties, and an assessment of the current level of understanding. It has been used widely in the scientific and policy communities. However, it has some important limitations, including the following:

- The figure does not provide information about the timescales over which each of the forcings is active. For example, the greenhouse gases in the first bar (CO_2, CH_4, N_2O, and halocarbons) remain in the atmosphere for decades or longer, whereas the various aerosols persist for days to weeks.
- The figure shows globally averaged forcings and therefore does not provide information about regional variation in forcing or vertical partitioning of forcing.
- The figure does not provide information about other climate effects of each forcing agent, such as impacts on the hydrological cycle.
- The figure gives the impression that one can simply sum the bars to determine an overall or net radiative forcing; however, such a calculation does not give a reasonable description of the cumulative effect of all the forcings.
- The uncertainty ranges are generally estimated from the range of published values and cannot be readily combined to determine a cumulative uncertainty.
- The figure does not consistently indicate the forcing associated with specific sources (e.g., coal, gas, agricultural practices).
- The figure omits nonradiative forcings as discussed in this report.

Although it would be unrealistic to expect a single figure to fully portray all of these aspects of radiative forcings, there are clearly opportunities to improve upon Figure 2-1 and to introduce new figures that address these limitations in the next IPCC report.

WELL-MIXED GREENHOUSE GASES

The radiative forcing due to CO_2, CH_4, N_2O, and various halocarbons is due to absorption of infrared (IR) radiation. It is well characterized and well understood. These gases remain in the atmosphere long enough to be well mixed; thus, their abundances are well known and have little spatial variability. Their concentrations have increased substantially since preindustrial times (see Table 2-1), and they are the greatest contributors to total anthropogenic radiative forcing. As shown in Figure 2-1, the IPCC estimate of the radiative forcing due to well-mixed greenhouse gases is +2.43 W m^{-2} from 1750 to 1998 (present), comprising CO_2 (1.46 W m^{-2}),

TABLE 2-1 Well-Mixed Greenhouse Gases

	CO_2	CH_4	N_2O	CFC-11	HFC-23	CF_4
Preindustrial concentration	~280 ppm	~700 ppb	~270 ppb	0	0	40 ppt
Concentration in 1998	365 ppm	1745 ppb	314 ppb	268 ppt	14 ppt	80 ppt
Rate of concentration change[a]	1.5 ppm/yr[b]	7.0 ppb/yr[d]	0.8 ppb/yr	-1.4 ppt/yr	0.55 ppt/yr	1 ppt/yr
Atmospheric lifetime	5 to 200 yr[c]	12 yr[d]	114 yr[d]	45 yr	260 yr	>50,000 yr

NOTE: CF_4 = perfluoromethane; CFC-11 = chlorofluorocarbon-11; HFC-23 = hydrofluorocarbon-23; ppm = parts per million; ppb = parts per billion; ppt = parts per trillion.

[a]Rate is calculated over the period 1990 to 1999.

[b]Rate fluctuated between 0.9 and 2.8 ppm yr^{-1} for CO_2 and between 0 and 13 ppb yr^{-1} for CH_4 over the period 1990 to 1999.

[c]No single lifetime can be defined for CO_2 because of coupling with surface reservoirs.

[d]This lifetime has been defined as an "adjustment time" that takes into account the indirect effect of the gas on its own residence time.

SOURCE: IPCC (2001).

CH_4 (0.48 W m^{-2}), N_2O (0.15 W m^{-2}), and halocarbons (0.34 W m^{-2}) (IPCC, 2001). The estimated uncertainty associated with this forcing is 10 percent, with that for CO_2 and N_2O being less and that for the other gases being greater. The estimated uncertainty for halocarbons is 10-15 percent for those molecules that have been studied in detail and is not well characterized for other halocarbons. Recent research on well-mixed greenhouse gases focuses on refining the models used to do radiative transfer calculations (e.g., Evans and Puckrin, 1999), considering the small temporal and spatial variations in concentrations which can lead to errors up to about 5-10 percent (Forster et al., 1997; Myhre and Stordal, 1997; Freckleton et al., 1998), and accounting for the extent to which clouds reduce radiative forcing (e.g., Myhre and Stordal, 1997).

The IPCC estimate for CH_4 forcing includes an observation-based estimate for both the direct forcing of CH_4 and the indirect forcing due to changes in the hydroxyl radical (OH) and tropospheric ozone (O_3) resulting from methane oxidation. The oxidation of CH_4 leads to a net loss of OH in the atmosphere, thereby lengthening the CH_4 lifetime. It is estimated that this indirect effect of CH_4 increases its radiative forcing by 25-35 percent over the direct CH_4 forcing (Lelieveld and Crutzen, 1992; Brühl, 1993; Lelieveld et al., 1993, 1998; Hauglustaine et al., 1994; Fuglestvedt et al., 1996). The oxidation of CH_4 also leads to the formation of tropospheric ozone, indirectly increasing the CH_4 forcing by 30-40 percent through the greenhouse effect of the additional tropospheric O_3. In the stratosphere, oxidation of CH_4 is a source of water vapor. In situ measurements of water vapor in the lower stratosphere indicate an increase of about 1 percent per year for 1954-2000 (Rosenlof et al., 2001), whereas satellite measurements of water vapor in the stratosphere in the 1990s showed no steady rate of change (Randel et al., 1996). A 1 percent annual increase in stratospheric water vapor would be associated with an estimated radiative forcing of 0.2 W m^{-2} since 1980 (Forster and Shine, 1999). The oxidation of CH_4 can explain only a fraction of such a water vapor increase.

TROPOSPHERIC AND STRATOSPHERIC OZONE

Atmospheric ozone modifies the radiative budget of the Earth system by absorbing radiation both in the IR and in the ultraviolet (UV). It acts both as a radiative forcing agent and as a climate feedback. Ozone is produced and destroyed by solar UV radiation and by chemical reactions involving natural and anthropogenic gases. Changes in ozone driven by anthropogenic emissions represent a forcing. However, ozone concentrations also respond to changes in temperature and UV radiation, transport patterns, and natural emissions from lightning and vegetation; these responses represent climate feedbacks. In what follows, tropospheric and stratospheric ozone are discussed separately because they are produced by different mechanisms and have very different radiative implications.

Tropospheric Ozone

Only 10 percent of atmospheric ozone resides in the troposphere, but this small fraction is of particular importance for climate forcing. Tropospheric ozone directly affects the radiative budget of the troposphere, and pressure broadening allows ozone absorption lines in the troposphere to extend into otherwise optically thin regions of the spectrum.

Ozone is produced in the troposphere by photochemical oxidation of volatile organic compounds (VOCs) and carbon monoxide (CO) in the presence of nitrogen oxides (NO_x = NO + NO_2). Anthropogenic emissions of these precursors have caused large increases in tropospheric ozone over the past century. The increase is estimated to be 50-100 percent globally according to current global three dimensional chemical transport models (CTMs), and the resulting radiative forcing is in the range 0.2-0.5 W m^{-2} (IPCC, 2001). An important caveat is that the CTMs are unable to reproduce the low ozone concentrations observed in the late nineteenth century and early twentieth century, suggesting (if the observations are correct) that they overestimate the natural source of ozone. Model calculations constrained with the historical observations indicate a larger forcing from anthropogenic tropospheric ozone, up to 0.8 W m^{-2} (Mickley et al., 2001; Shindell and Faluvegi, 2002).

Beyond this direct radiative forcing effect, tropospheric ozone also has an indirect effect as the primary precursor of the OH radical. Increasing ozone causes tropospheric OH to increase, thus decreasing the lifetime of methane and facilitating aerosol nucleation (both negative forcings). Assessing this indirect effect is complicated because the increase in ozone is driven by emissions of its precursors, which themselves have intrinsic effects on OH. Increasing NO_x thus causes OH to increase, while increasing CO and VOCs cause OH to decrease. According to the current generation of CTMs, OH concentrations have decreased by about 10 percent over the past century (Wang and Jacob, 1998). Another indirect effect of tropospheric ozone is to cool the stratosphere (Joshi et al., 2003; Mickley et al., 2004), affecting stratospheric ozone and polar stratospheric cloud (PSC) levels.

The greatest uncertainty in quantifying the direct radiative forcing from tropospheric ozone lies in reconstructing its concentration field in the past and projecting it into the future. The inability of current models to reproduce ozone observations from the early twentieth century could reflect calibration problems in the observations, as well as model errors in the estimates of natural sources. CTMs also have problems in simulating the well-calibrated ozone trends over the past 30 years (Fusco and Logan, 2003), implying that fundamental problems remain in our understanding of tropospheric ozone chemistry. Uncertainty in quantifying the indirect radiative effect of ozone is related mainly to the complexity of factors controlling OH concentrations (Lawrence et al., 2001). Uncertainties in predicting the climatic response to changes in tropospheric ozone are also large and require further investigation using general circulation models (GCMs).

Stratospheric Ozone

Depletion of stratospheric ozone over the past 30 years has caused both a positive radiative forcing at the Earth's surface (due to increased UV penetration) and a negative forcing (due to reduced IR emission from the stratosphere to the troposphere). The consensus from current radiative models constrained by observed ozone trends is that the net forcing is negative and of magnitude -0.10 ± 0.05 W m^{-2}. Forster and Tourpali (2001) argue that about half of this forcing is due to an increase in tropopause heights and thus should not be considered a forcing but rather a feedback. The main indirect radiative effects of stratospheric ozone depletion are (1) increased UV penetration to the troposphere, increasing tropospheric OH concentrations and hence decreasing the lifetime of methane (IPCC, 2001), and (2) changes in stratospheric water vapor.

Several GCM studies have examined the climate response to changes in stratospheric ozone. Shindell et al. (1999) finds that changes in the upper stratosphere elicit far greater surface climate response than changes in the lower stratosphere. Stuber et al. (2001) find that changes in stratospheric ozone have a greater effect per unit forcing than changes in CO_2, largely because of feedbacks associated with stratospheric water vapor.

The greatest uncertainty in quantifying radiative forcing from past changes in stratospheric ozone is the vertical distribution of the ozone trend in relation to temperature, since the magnitude of the forcing depends crucially on temperature (IPCC, 2001). Another critical issue is to better quantify indirect radiative forcings, particularly the effect on stratospheric water vapor, which could double the effective forcing according to Stuber et al. (2001).

DIRECT EFFECT OF AEROSOLS

Aerosol particles both scatter and absorb radiation, representing a direct radiative forcing; scattering generally dominates (except for black carbon particles) so that the net effect is of cooling. Global models have demonstrated the important role of sulfate aerosols in providing the cooling effect missing in past models of the atmospheric radiation balance (Kiehl et al., 1995). The average global mean aerosol direct forcing from fossil fuel combustion and biomass burning is in the range from -0.2 to -2.0 W m^{-2} (IPCC, 2001). This large range results from uncertainties in aerosol sources, composition, and properties used in different models. Recent advances in modeling and measurements have provided important constraints on the direct effect of aerosols on radiation (Ramanathan et al., 2001a; Russell et al., 1999; Conant et al., 2003). Critical gaps, discussed further below, relate to

- spatial heterogeneity of the aerosol distribution, which results from the short lifetime (a few days to a week) against wet deposition;
- chemical composition, especially the organic fraction;
- mixing state and behavior (hygroscopicity, density, reactivity, and acidity); and
- optical properties associated with mixing and morphology (refractive index, shape, solid inclusions).

The chemical composition of particles is in general not well known. The mixing state and relative humidity history of sulfate-nitrate-ammonium aerosols have important implications for their water content and hence their direct radiative effect (Martin et al., 2004). Uncertainties are particularly large for the 20 to 70 percent of particle mass that consists of organic compounds (NARSTO, 2003). Measurement of organic components is inherently difficult for three reasons: (1) small sample sizes and analytical difficulties, (2) the complexity of mixtures, and (3) artifacts in sampling procedures. The ideal approach for characterizing organic mass in aerosol particles would identify, molecule by molecule, the composition of each individual particle. Such instrumentation is unlikely to become available in the near future. In the meantime, it will be important to use partial information from traditional and new approaches, including evolved gas analysis techniques, gas chromatography, time-of-flight and chemical ionization mass spectrometry, Fourier transform infrared spectroscopy, and near-edge X-ray absorption fine structure (Morrical et al., 1998; Schauer et al., 1999; Huebert and Charlson, 2000; Russell et al., 2002; Bahreini et al., 2003; Russell, 2003).

Radiative and climate models generally assume that aerosols are "externally mixed," that is, individual particles are made up of a single component (Koch, 2001; Cooke et al., 2002). Actual aerosol particles are multicomponent mixtures for which properties, such as water uptake, differ from those expected from the simple addition of components because of nonlinear interactions between components. The consequence for radiative forcing is that water uptake by particles is not predicted accurately. The presence of organic compounds has two competing effects on particle hygroscopicity: (1) it reduces the mass of water taken up, and (2) it initiates water uptake at a lower relative humidity (called the deliquescence relative humidity) (Ming and Russell, 2002). The reduction in water uptake results from the typically low solubility of organic compounds. Particles containing organic compounds will grow less as a function of relative humidity, meaning that models that use the properties of sulfate aerosol will overestimate the direct radiative forcing.

Many GCM aerosol schemes tend to omit organic particles or underestimate their size at typical boundary layer humidity (~80 percent). Chung and Seinfeld (2002) estimate the effect of organic carbon on radiative forcing of the climate to be between –0.09 and –0.21 W m^{-2}, with the range of uncertainty driven by the role of water uptake by organic aerosols. The combination of the role of organic carbon and its water uptake with the externally or internally mixed states of other components results in a direct aerosol forcing range of –0.86 to –1.26 W m^{-2}, or an uncertainty of ±50 percent (Chung and Seinfeld, 2002). In addition to water uptake, uncertainties in aerosol lifetime and optical properties contribute to the range of uncertainty.

In addition to these general difficulties in describing the direct radiative forcing from aerosols, specific uncertainties relate to (1) light-absorbing black carbon, and (2) categorization of aerosol types in modeling. These are discussed below.

Black Carbon

Individual aerosol particles may contain light-absorbing carbon-containing compounds referred to collectively as "black" carbon (BC) or equivalently as soot. In addition to elemental carbon, BC frequently includes low-volatility solid or liquid organic compounds, typically composed of long hydrocarbon chains with high molecular weights (Marley et al., 2001). The presence of trace amounts of BC (as little as 5 to 10 percent of the total mass in anthropogenic aerosols) can result in large atmospheric solar absorption. This absorption can be enhanced when BC is embedded in refractive particles (Chylek et al., 1996; Fuller et al., 1999). Current understanding of the global emission of BC is uncertain by factors of two or more (Cooke et al., 2002; Bond et al., 2004). Biomass burning and fuel combustion are the two main contributors. BC has been detected in remote oceanic regions, implying hemispheric-wide dispersal.

The direct forcing due to carbonaceous aerosols can be separated into three components:

1. A portion of the direct solar beam is scattered back to space, which leads to a reduction in solar radiation reaching the surface. This reduction

manifests as increased reflection at the top of the atmosphere (TOA), i.e., a negative radiative forcing (cooling).

2. A portion of the direct solar beam is absorbed by the aerosol, and this atmospheric absorption leads to further reduction in solar radiation reaching the surface. As shown later, this shielding of the surface by BC is the dominant absorption term for anthropogenic aerosols with as little as 10 percent of BC. This absorption leads to a *positive* radiative forcing of the atmosphere and a *negative* radiative forcing of the surface.

3. The upward diffuse beam from scattered radiation is absorbed by the BC aerosol, reducing the solar radiation that escapes to space and resulting in a *positive* radiative forcing for the surface-atmosphere column. This effect could be large in cloudy skies if BC lies above low clouds (Haywood and Ramaswamy, 1998).

The TOA forcing reported in IPCC and other global warming studies is the sum of processes (1) and (2). Process (3), albeit the largest in terms of magnitude, does not contribute significantly to TOA forcing, since it adds solar radiation to the atmosphere and reduces surface solar heating by the same magnitude. The net effect of BC is to increase the radiative heating of the atmosphere and decrease the radiative heating of the surface. The TOA radiative forcing is the sum of the surface and the atmospheric forcing. At the TOA the BC effect opposes the cooling effect of sulfates and organics, while at the surface all aerosols lead to reduction of solar radiation. Thus, aerosol-induced changes at the surface can far exceed those at the TOA.

The direct effect of BC aerosol in the atmosphere has important implications for "global dimming." Black carbon emissions may have increased by a factor of two to four during the last 50 years (Novakov et al., 2003). Given such large increases in BC emissions and the large impact of BC on reducing surface solar radiation, large decreases in surface solar radiation should be observed downwind of major sources of BC. Long-term negative trends in surface solar irradiances have been observed by surface radiometers worldwide (Ohmura et al., 1998; Stanhill and Cohen, 2001; Liepert, 2002). The reported trends in the annual mean irradiance vary from −5 percent (10 W m^{-2}) between 1958 and 1985 for all land stations to about −1 to −3 percent per decade for the last four decades over many of the 1500 stations in the global datasets. The decreases are so large that there is skepticism about the measurements. Trends in surface radiometer all-sky observations are subject to large uncertainties due to difficulties in maintaining accurate calibration for routine surface observations in remote locations and measurement errors inherent in broadband radiometric measurements (Dutton et al., 2001).

This global dimming is thought to be caused by both absorbing aerosols and increases in cloud cover (Liepert, 2002). Most of the radiometer observations are over land, and we need to understand whether they are affected by local urban haze. Anthropogenic absorbing aerosols, by themselves, can reduce land-averaged solar radiation by about 3 to 5 W m^{-2} (Ramanathan et al., 1995; Jacobson, 2002). Such large reductions in surface solar radiation have implications for the hydrological cycle, since roughly 70 percent of the absorbed solar radiation is balanced by the latent heat flux of evaporation.

This global dimming may also be related to changes in the ratio of direct and diffuse solar irradiance received by vegetation. As documented by Gu et al. (2003), the increase in diffuse irradiance for the two years following the eruption of Mount Pinatubo in 1991 resulted in a 23 percent increase of noontime photosynthesis of a deciduous forest in 1992 under cloudless conditions. The increased diffuse irradiance permits a greater penetration of photosynthetically active sunlight into the canopy. However, if there is a sufficient reduction of total solar irradiance received at the ground, vegetation growth could be stunted. Chameides et al. (1999), for example, reported on reductions in crop yield in China due to reduced total solar irradiance from pollution aerosols. Krakauer and Randerson (2003) concluded from tree ring data that with respect to volcanic eruptions, the beneficial effect of aerosol light scattering for high northern latitudes appears to be offset entirely by the deleterious effect of eruption-induced climate change. Using field observations, Niyogi et al. (2004) found that increased aerosol loading led to increases in carbon assimilation for forests and crops and decreases for grasslands. The effect on carbon assimilation was larger with aerosols than with clouds, since clouds reduced the total solar irradiance more than the aerosols did.

Deposition of BC aerosols over snow-covered areas can result in changes to the surface albedo (Chylek et al., 1983). Further reductions in albedo occur due to the enhanced melting that accompanies the heating of absorbing soot particles in snow. Chylek et al. (1983) estimate this enhancement to be up to a factor of ten in the rate of melting. Recent model results indicate radiative forcings of +0.3 W m^{-2} in the Northern Hemisphere associated with albedo effects of soot on snow and ice (Hansen and Nazarenko, 2004).

Model Discretization of Aerosols

Typically the behavior of aerosol particles in the atmosphere has been described in models by discretization of both size and chemical composition. The continuous particle size spectrum is described by a limited number of modes, moments, or sections (Seinfeld and Pandis, 1998). For many problems, such as the evolution of marine aerosol, the computational simplicity of a few modes is sufficient to characterize changes in the particle distribution (Pandis et al., 1994). Other problems, including those involving aerosol-cloud interactions, require ~100 size categories to accurately represent the indirect effect of particles on cloud properties (Russell et al., 1999). While the simplified models are known to represent measured size distributions incompletely, the uncertainties associated with these simplifications can easily be quantified. GCMs employing this approach have not yet had the computational power to evaluate the sensitivity of aerosol radiative forcing to their simplified treatments of aerosol size distributions.

Initial models of particle evolution assumed that the composition of particles consists of a single internal mixture, both for computational simplicity and for lack of contradictory measurements (Warren and Seinfeld, 1985). In the last decade, chemical composition has been discretized in a fashion similar to the treatment of particle size, typically by "lumping" particles with similar chemical compositions into a few categories. This approach has been used to represent the externally mixed nature of aerosol particles, typically including categories such as "sulfate," "sea salt," "dust," and "carbonaceous" (Jacobson et al., 1994; Pandis et al., 1994; Russell and Seinfeld, 1998; Jacobson, 2001; Koch, 2001; Garrett et al., 2003). The small number of categories has enabled their inclusion in GCM simulations. However, this does not reflect the variety of mixtures actually present in the troposphere (Murphy et al., 1997). The complexity of particle structures, their heterogeneities, and their mixing states (Russell et al., 2002) will have to be addressed to represent their hygroscopic and optical properties. For example, a small amount of absorbing organic compounds mixed in with sea salt aerosol can be sufficient to change the associated radiative forcing from negative to positive, especially over low albedo surfaces such as clouds, ice, and coastal areas (Randles et al., 2004).

INDIRECT EFFECT OF AEROSOLS

Aerosols interact with clouds in several ways that can affect the climate system, in particular by acting as cloud condensation nuclei (CCN) and ice nuclei. These interactions are generally referred to as the indirect effects of aerosols. Table 2-2 summarizes the various aerosol indirect effects. As shown in the table, aerosols can lead to both warming (positive forcing) and cooling (negative forcing), complicating the analysis of their net effect.

Aerosols have several indirect effects on warm stratiform clouds. The Twomey effect, also known as the first indirect aerosol effect, refers to the enhanced reflection of solar radiation due to more but smaller cloud droplets in a cloud whose liquid water content remains constant (Twomey, 1959). The IPCC Third Assessment Report concluded that the first indirect aerosol effect amounts to 0 to -2 W m^{-2} in the global mean (IPCC, 2001). In addition, more but smaller cloud droplets reduce the precipitation effi-

TABLE 2-2 Overview of the Different Aerosol Indirect Effects Associated with Clouds

Effect	Cloud Type	Description	Sign of TOA Radiative Forcing
First indirect aerosol effect (cloud albedo or Twomey effect)	All clouds	For the same cloud water or ice content, more but smaller cloud particles reflect more solar radiation	Negative
Second indirect aerosol effect (cloud lifetime or Albrecht effect)	All clouds	Smaller cloud particles decrease the precipitation efficiency, thereby prolonging cloud lifetime	Negative
Semidirect effect	All clouds	Absorption of solar radiation by soot leads to evaporation of cloud particles	Positive
Glaciation indirect effect	Mixed-phase clouds	An increase in ice nuclei increases the precipitation efficiency	Positive
Thermodynamic effect	Mixed-phase clouds	Smaller cloud droplets inhibit freezing, causing supercooled droplets to extend to colder temperatures	Unknown
Surface energy budget effect	All clouds	The aerosol-induced increase in cloud optical thickness decreases the amount of solar radiation reaching the surface, changing the surface energy budget	Negative

ciency and therefore enhance cloud lifetime and, hence, cloud reflectivity, which is referred to as the second indirect aerosol or cloud lifetime effect (Albrecht, 1989). Absorption of solar radiation by black carbon leads to heating of the air, which can result in an evaporation of cloud droplets. This is referred to as the semidirect effect (Hansen et al., 1997). The absorption of soot is 2 to 2.5 times greater if soot is present in the cloud droplets (Chylek et al., 1996); thus the magnitude of the semidirect effect depends crucially on the location of black carbon with respect to the cloud. This warming can partially offset cooling due to the indirect aerosol effect. Both the cloud lifetime effect and the semidirect effect involve feedbacks because the cloud lifetime and cloud liquid water content change; therefore, they were not included in the radiative forcing bar chart of the IPCC (2001) assessment.

The committee notes that separating the aerosol indirect forcing into "first" and "second" kinds is not necessarily a useful construct. Although the first kind accounts for a significant and distinguishable set of properties between polluted and unpolluted clouds, the lifetimes of clouds are such that it is never really observed. Clouds respond quickly to shifts in droplet sizes, so by the time observations are made, clouds have already progressed into the "second" mode. Alternatively, the indirect forcing could be calculated for clouds with fixed water amounts or for clouds with water amounts that are free to adjust to the changes in droplet sizes.

Surface-based and satellite data have provided evidence of these aerosol effects on warm clouds. Feingold et al. (2003) used various observations at the Department of Energy Atmospheric Radiation Measurement (ARM) site in Oklahoma to estimate the indirect aerosol effect from the partial derivative of the logarithm of cloud droplet radius with respect to the logarithm of the aerosol extinction. Defined in this way, the indirect aerosol effect can be compared to its estimates from satellite retrievals. They find that the indirect effect at the ARM site is larger than estimated from POLDER satellite data by Breon et al. (2002). However, Rosenfeld and Feingold (2003) pointed out that limitations of the POLDER satellite retrievals could explain the discrepancy. Penner et al. (2003) combined ARM data together with a Lagrangian parcel model at the ARM sites in Oklahoma and Alaska to provide observational evidence of a change in radiative forcing due to the indirect aerosol effect. Long-term observations with satellites over Europe and China show evidence for the semidirect effect (Krüger and Graßl, 2002, 2004). Kim et al. (2003) reported regional aerosol influences on TOA flux of up to 50 W m^{-2}. Schwartz et al. (2002) reported albedo enhancements due to the Twomey effect of as much as 0.2. None of these techniques, however, enable estimates of the anthropogenic indirect aerosol effect globally.

Climate model estimates of the first and second indirect aerosol effects and the semidirect aerosol effect are still very uncertain. Kristjansson (2002) concluded that the Twomey effect is three times more important than the cloud lifetime effect, whereas Lohmann et al. (2000) simulated a cloud lifetime effect that is 40 percent larger than the Twomey effect. Estimated magnitudes of the Twomey effect and cloud lifetime effect are -0.5 to -1.9 W m^{-2} and -0.3 to -1.4 W m^{-2}, respectively (Lohmann and Feichter, 2004). Based on model results by Penner et al. (2003) and Lohmann and Feichter (2001), the estimated magnitude of the semidirect effect is $+0.1$ to -0.5 W m^{-2}. Whereas climate models predict an increase in liquid water due to the cloud lifetime effect, observations of ship tracks show that polluted clouds have less liquid water, not more (Platnick et al., 2000; Coakley and Walsh, 2002). Thus, the enhancement of cloud water through the suppression of precipitation in polluted clouds is still not understood completely and therefore contributes to the significant uncertainty attributed to the aerosol indirect forcing.

Liu and Daum (2002) estimated that the magnitude of the first indirect aerosol effect can be reduced by 10 to 80 percent by including the influence that an increasing number of cloud droplets has on the shape of the cloud droplet spectrum. When this dispersion effect is taken into account in global climate models, the reduction is rather moderate and amounts to 15 to 35 percent (Peng and Lohmann, 2003; Rotstayn and Liu, 2003). Lower estimates of the indirect aerosol effect are in better agreement with inverse calculations based on historical climate record data of oceanic and atmospheric warming (Forest et al., 2002; Knutti et al., 2002; Anderson et al., 2003a).

The presence of ice in large-scale or convective clouds allows for further aerosol-cloud interactions. For large-scale, mixed-phase clouds,

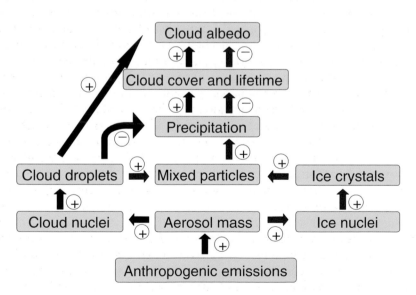

FIGURE 2-2 Schematic of the warm indirect aerosol effect (blue arrows) and glaciation indirect aerosol effect (red arrows). SOURCE: Adapted from Lohmann (2002).

Lohmann (2002) showed that if in addition to mineral dust, a fraction of the hydrophilic soot aerosol particles is assumed to act as contact ice nuclei at temperatures between 0° and −35°C, then a "glaciation indirect effect" results (Figure 2-2). Increases in contact ice nuclei in the present-day climate result in more frequent glaciation of clouds and increase the amount of precipitation via the ice phase. Observations in the presence of Saharan African dust indicate that mildly supercooled clouds at temperatures between −5 and −9°C are already glaciated (Sassen et al., 2003).

For convective mixed-phase clouds, Rosenfeld and Woodley (2000) analyzed aircraft data together with satellite data to show that pollution aerosols suppress precipitation. This hypothesis is supported by a modeling study with a cloud-resolving model by Khain et al. (2001). Taking these results to the global scale, Nober et al. (2003) evaluated the sensitivity of the general circulation to the suppression of precipitation by anthropogenic aerosols by implementing a simple warm cloud microphysics scheme into convective clouds. They found large instantaneous local aerosol forcings reducing the warm-phase precipitation (thermodynamic effects). Menon et al. (2002b) showed that absorbing aerosols over China change the atmospheric stability and vertical motion by heating the air and, thus, the large-scale circulation and the hydrological cycle.

The IPCC aviation report (Penner et al., 1999) identified the effects of aircraft on upper tropospheric cirrus clouds as a potentially important climate forcing. One aspect may be described as the direct effect due to the formation of condensation contrails as a result of supersaturated air from the aircraft. This effect can be nonnegligible as was found during the three-day grounding of all U.S. commercial aircraft following the September 11, 2001, terrorist attacks. An anomalous increase in the average diurnal temperature range over the United States was observed and partly attributed to the absence of contrails from jet aircraft (Travis et al., 2002). Aerosols emitted by the aircraft may also cause indirect effects associated with an increase in ice nuclei in the upper troposphere. Evidence of a climate effect of air traffic was first provided by Boucher (1999) who used ship-based measurements of cloud cover together with fossil fuel consumption data for aircraft to show that recent increases in air traffic fuel consumption are accompanied by an increase in cirrus cloudiness. Recent studies (Lohmann and Kärcher, 2002) suggest that the impact of aircraft sulfur emissions on cirrus properties via homogeneous freezing of sulfate aerosols is probably small. Hence the question has been raised whether aircraft-generated black carbon particles serving as heterogeneous ice nuclei, as found by Ström and Ohlsson (1998), may have a significant impact on cirrus cloudiness and cirrus microphysical properties.

Aerosols can also modify latent and sensible heat fluxes at the surface, thus exerting a nonradiative forcing on the hydrological cycle. Increasing aerosol and cloud optical depth cause a reduction of solar radiation at the surface. For the surface energy balance to reach a new equilibrium state, the latent and sensible fluxes have to adjust. Because evaporation has to equal precipitation on the global scale, a reduction in the latent heat flux leads to a reduction in precipitation. As shown in model simulations by Liepert et al. (2004), despite an increase in greenhouse gases, increases in optical depth due to the direct and indirect anthropogenic aerosol effects can cause a reduction in evaporation and precipitation. This mechanism is consistent with observations of decreased evaporation from open pans of water over the last 50 years. Roderick and Farquhar (2002) characterized steadily decreasing pan evaporation due to decreases in solar irradiance resulting from increasing cloud coverage and aerosol concentration. Increasing aerosol optical depth associated with scattering aerosols alone in otherwise clear skies has been shown to produce a larger fraction of diffuse radiation at the surface, which results in greater carbon assimilation into vegetation (and therefore greater transpiration) without a substantial reduction in the total surface solar radiation (Niyogi et al., 2004).

LAND-COVER AND LAND-USE CHANGES

Land-use changes include irrigation, urbanization, deforestation, desertification, reforestation, grazing of domestic animals, and dryland farming. Each of these alterations in landscape produces significant changes in radiative forcing (e.g., Pitman, 2003; Kabat et al., 2004). Global maps of land-cover changes over the past 300 years are shown in Figure 2-3. In addition, changes in tropical forests have been reported by O'Brien (2001). There are historical land use datasets (e.g., Ramankutty and Foley, 1999) as well as satellite-based land-cover datasets. Satellite products include the DISCover dataset developed under the auspices of the International Geosphere-Biosphere Programme (Loveland et al., 2000) and land-cover datasets based on MODIS/Terra data (Strahler et al., 1999).

Land-use and land-cover changes can affect the Earth's radiative bal-

---

FIGURE 2-3 Global estimate of land use and land cover for (a) 1700, (b) 1900, and (c) 1990. The human-disturbed landscape includes intensive cropland (red) and marginal cropland used for grazing (pink). Other landscape includes, for example, tropical evergreen and deciduous forest (dark green), savannah (light green), grassland and steppe (yellow), open shrubland (maroon), temperate deciduous forest (blue), temperate needleleaf evergreen forest (light yellow), and hot desert (orange). SOURCE: Klein Goldewijk (2001).

Year 1700

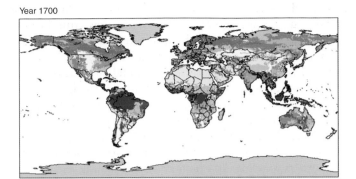

Year 1900

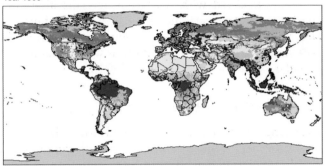

Year 1990

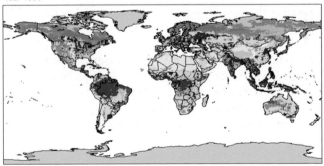

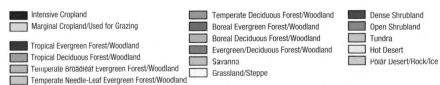

Intensive Cropland
Marginal Cropland/Used for Grazing

Tropical Evergreen Forest/Woodland
Tropical Deciduous Forest/Woodland
Temperate Broadleaf Evergreen Forest/Woodland
Temperate Needle-Leaf Evergreen Forest/Woodland

Temperate Deciduous Forest/Woodland
Boreal Evergreen Forest/Woodland
Boreal Deciduous Forest/Woodland
Evergreen/Deciduous Forest/Woodland
Savanna
Grassland/Steppe

Dense Shrubland
Open Shrubland
Tundra
Hot Desert
Polar Desert/Rock/Ice

ance both directly and indirectly. Direct effects include the change of albedo and emissivity resulting from the different types of land covers that modify the amount of shortwave radiation absorbed at the surface and of longwave radiation absorbed and emitted at the surface. For example, the development of agriculture in tropical regions typically results in an increase of albedo from a low value of forest canopies (0.05-0.15) to a higher value of agricultural fields, such as pasture (0.15-0.20). In contrast, irrigated fields in arid areas tend to have a lower albedo than the bare, typically bright soils they cover. The seasonal variation of albedo as a result of land-cover change can also have pronounced effects on the net radiation at the Earth's surface.

As shown in Figure 2-1, the IPCC (2001) reports the global-averaged forcing due to albedo change alone as -0.25 ± 0.25 W m^{-2}. The level of scientific understanding is listed as "very low." The uncertainties in the albedo change reflect the complexity of the land surface (e.g., type of vegetation, phenology, density of coverage, soil color). When aggregating regional information about land surface up to the global scale, large global average uncertainty ranges result. A recent assessment of the albedo change estimates a range of -0.6 to 0.5 W m^{-2}, with the negative values being more likely (Myhre and Myhre, 2003).

Indirect effects of land-cover change on the net radiation include a variety of processes related to (1) the ability of the land cover to use the radiation absorbed at the ground surface for evaporation, transpiration, and sensible heat fluxes (the impact on these heat fluxes caused by changes in land cover is sometimes referred to as *thermodynamic forcing*); (2) the exchange of greenhouse and other trace gases between the surface and the atmosphere; (3) the emission of aerosols (e.g., from dust); and (4) the distribution and melting of snow and ice. These effects are discussed below.

Indirect Effects of Land-Use and Land-Cover Change

Changes in soil wetness can significantly modify the energy balance of continental surfaces. When soil moisture is high, most of the radiative energy absorbed at the ground surface is used for physical evaporation and transpiration of water. The latent heat flux is large, the sensible heat flux is small (in arid areas, it can even be negative, a process known as the "oasis effect"), the land-surface temperature is relatively low (compared to conditions with more sensible heat flux with the same net radiation), and as a result, the longwave radiation emitted by the land surface is relatively low. As a result, the atmospheric boundary layer that develops above such land is typically thin and moist. In contrast, when the soil is dry, there is no latent heat flux, the sensible heat flux is large, the land surface temperature is higher, and as a result, the longwave radiation emitted by the land surface is relatively high. The planetary boundary layer developing above such land is typically deep and dry. Note also that soil wetness is a function of precipitation, which itself can be affected by land cover, resulting in complex land-atmosphere feedbacks. For example, Figure 2-4 illustrates schematically the alteration of fluxes as a result of the conversion of forest to cropland. The conversion alters the transpiration of water vapor into the atmosphere. The surface albedo is usually higher when vegetation is reduced, exposing some of the soil.

Land-use and land-cover change can also have indirect effects by affecting fluxes of greenhouse gases from the land surface. Land cover influences the release into the atmosphere of water vapor, a greenhouse gas with obvious impacts on clouds and precipitation. The biological activity at the ground surface can be a sink or a source of carbon dioxide, methane, and other gases, but these processes are still not understood and are not represented well (if at all) in climate models.

Interactions between the land and atmosphere complicate the interpretation of CO_2 trends. CO_2 concentrations change in response to vegetation, ocean, and erosional feedbacks, as well as anthropogenic emissions; some of these fluxes are responses to variations of the climate system on multiple

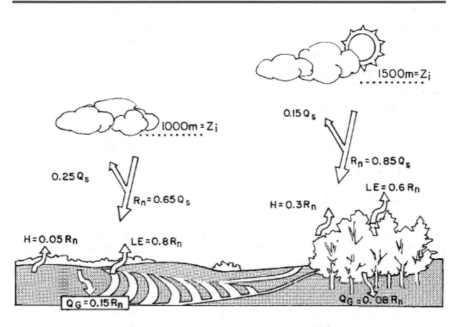

FIGURE 2-4 Schematic, based on observations in southwest France, of the influence on the surface energy budget of land-use change from forest to cropland. SOURCE: Kabat et al. (2004).

timescales. Biophysical feedbacks (such as transpiration) operate on time-scales of minutes, while biogeochemical feedbacks (such as plant growth) become significant only after several days and longer. Biogeographic changes (such as species composition) occur over years and longer.

Land cover affects the amount of aerosols that can be lifted by wind into the atmosphere. In general, the less dense the vegetation and the more intense the human activity on the land, the more aerosols can be lifted by dust storms and other updrafts. The impact of aerosols is discussed in detail in other sections of this report.

Snow has a very significant impact on the land-cover albedo and, as a result, on the radiation balance of the land cover. However, the same amount of snow at a specific location and time has a different synergistic effect with different land-cover types. For example, the albedo of an evergreen coniferous tree canopy is lower than that of an adjacent snow-covered clearing. Land type also affects the duration of snow on the ground.

Partitioning Between Latent and Sensible Heat Fluxes on a Regional Scale

Changes in the partitioning of net radiative fluxes into sensible and latent can substantially alter the atmospheric circulation. Using the results from Chase et al. (2000a), where a conservative estimate of land-cover change by humans was specified, Pielke et al. (2002) reported a globally averaged redistribution of sensible and latent turbulent heat fluxes on the order of 1 W m^{-2}. Therefore, the spatial redistribution of the surface turbulent fluxes indicates that the net radiation received at the surface is changed in how the energy is inserted into the atmosphere. The surface forcing of climate, as a result, is altered.

The effect on local climate can be substantial. Marshall et al. (2004a,b) documented major alterations in summer rainfall and temperature, and in freeze occurrence, due to land conversion in the twentieth century in Florida. The observed land change between the pre-1900 period and 1993 in Florida is shown in Figure 2-5. Figure 2-6 shows model results for the change in net radiation at the surface, averaged over July-August, in response to that land-use change. Over the land area shown in Figure 2-5, the two-month, area-averaged reduction in rainfall was 10-12 percent. Freeze intensity and duration also changed as a result of land-use conversion. In the agricultural area just south of Lake Okeechobee, the draining of marshes and their replacement with orchards and other agricultural crops resulted in greater radiation loss to space with a resultant longer and more severe freeze.

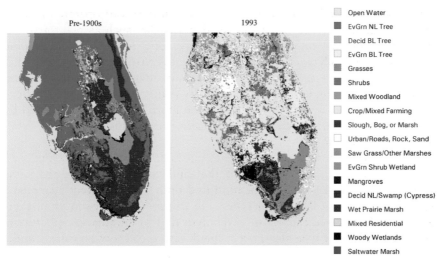

	☐ Open Water
Pre-1900s 1993	■ EvGrn NL Tree
	■ Decid BL Tree
	☐ EvGrn BL Tree
	■ Grasses
	■ Shrubs
	■ Mixed Woodland
	☐ Crop/Mixed Farming
	■ Slough, Bog, or Marsh
	☐ Urban/Roads, Rock, Sand
	■ Saw Grass/Other Marshes
	■ EvGrn Shrub Wetland
	■ Mangroves
	■ Decid NL/Swamp (Cypress)
	■ Wet Prairie Marsh
	☐ Mixed Residential
	■ Woody Wetlands
	■ Saltwater Marsh

FIGURE 2-5 U.S. Geological Survey land-cover data in Florida for (*left*) pre-1900 and (*right*) 1993. SOURCE: Marshall et al. (2004b).

SOLAR FORCING

Irradiance

The Sun's electromagnetic radiation powers the Earth's climate. The blackbody temperature of the solar surface is about 5800 K, and as a result, the spectrum of solar radiation peaks at visible wavelengths. Solar irradiance is the electromagnetic radiation from the Sun incident at the top of the Earth's atmosphere at a distance of one astronomical unit (1 AU = 1.5 × 10^{11} m), corresponding to the mean Earth-Sun distance. Total solar irradiance (TSI), the integral of spectral irradiance, has a mean value of 1365 ± 1 W m^{-2} according to Active Cavity Radiometer Irradiance Monitor (ACRIM), Earth Radiation Budget Satellite (ERBS), and Solar and Heliospheric Observatory (SOHO) space-based radiometry. The recently launched Solar Radiation and Climate Experiment (SORCE) radiometer suggests a lower absolute value of 1358 W m^{-2}. Approximately equal amounts of energy are radiated above and below 742 nm.

Solar irradiance forcing of climate arises from the change in electromagnetic radiation at 1 AU produced by variations in the Sun's activity. This forcing is the result of changes generated within the Sun, not changes in the Earth's orbit around the Sun. Calibrated radiometers on various

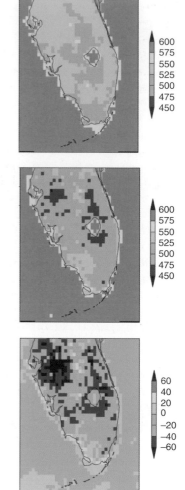

FIGURE 2-6 July-August average net daytime (12Z-23Z) radiation received at the surface for the pre-1900 landscape (*top*), 1989 large-scale meteorology with the 1993 landscape (*middle*), and the difference between the 1989 simulation and the pre-1900 case (*bottom*) in units of W m^{-2}. The daytime land average for the pre-1900 case is 530 W m^{-2} and for the 1989 simulation is 503 W m^{-2}. SOURCE: Adapted from Marshall et al. (2004b).

spacecraft have measured the total solar irradiance since the late 1970s. There is an 11-year cycle in total solar irradiance of peak-to-peak amplitude ~1 W m^{-2} (0.1 percent) in the past three cycles. Allowing for reflection of 30 percent of this incident energy (Earth's albedo) and averaging over the globe, the corresponding climate forcing is of order 0.2 W m^{-2}.

Multiple TSI datasets have been combined into a composite time series of daily total solar irradiance from 1979 to the present. This requires the cross-calibration of measurements made by overlapping datasets to adjust for the different absolute scales of individual radiometers and the comparison of time-dependent trends among radiometers having different solar exposures and different design. Different assumptions about radiometer performance led to different reconstructions of TSI for the past two decades. In one reconstruction (Fröhlich and Lean, 2002), an 11-year total irradiance cycle of amplitude ~1 W m^{-2} (0.1 percent) has approximately equal values during successive solar minima (1986 and 1996), whereas in another reconstruction, the 1996 minimum is 0.5 W m^{-2} (0.05 percent) higher than in 1986. Willson and Mordvinov (2003) argue that this is evidence for a long-term secular trend in total irradiance underlying the activity cycle. If real, this increase in total solar irradiance would imply a secular climate forcing of 0.1 W m^{-2} over the past two decades. The actual increase, however, occurred during a two-year interval from 1990 to 1992 and coincides with apparent drifts in the ERBS radiometer on the Nimbus 7 spacecraft, used in the reconstructions to connect observations by ACRIM I and II radiometers. This argues against any detectable long-term trend in the observed irradiance to date. Likewise, models of total solar irradiance variability that account for the influences of solar activity features—dark sunspots and bright faculae—do not predict a secular change in the past two decades.

Knowledge of solar irradiance variations is rudimentary prior to the commencement of continuous space-based irradiance observations in 1979. Models of sunspot and facular influences developed from the contemporary database have been used to extrapolate daily variations during the 11-year cycle back to about 1950 using contemporary sunspot and facular proxies, and with less certainty annually to 1610. The reconstructions are based on records of the number of sunspots on the Sun's surface, which commenced with the first telescopic observations of the Sun in the early seventeenth century. These remain the longest existing direct record of solar activity.

Most historical reconstructions of total solar irradiance have assumed that the 11-year activity cycle is superimposed on a longer-term varying background component. In these reconstructions, which have been widely used in climate change simulations, estimated TSI increases from the Maunder Minimum in the late seventeenth century to the present are in the range of 0.2 to 0.4 percent (Lean, 2000), which corresponds to climate forcing in the range of 0.4 to 0.6 W m^{-2}. Circumstantial evidence from cosmogenic isotope proxies of solar activity (^{14}C and ^{10}Be) and plausible variations in Sun-like stars motivated the assumption of long-term secular irradiance trends, but recent work questions the evidence from both. Preliminary modeling of magnetic field transport on the solar surface suggests that cosmogenic isotopes (which are controlled by heliospheric magnetic fields) can vary in different ways and have larger secular trends than irradiance (which is controlled by magnetic fields confined to the solar atmosphere)

(Lean et al., 2002). Critical examination of a broader distribution of stars and examination of their "solar-likeness" appear to contradict the initial findings that noncycling Sun-like stars undergo Maunder-type episodes with reduced overall brightness. Likewise, an initial study of (uncalibrated) solar images failed to find evidence of a varying brightness component in the past century (Foukal and Milano, 2001). The most recent studies therefore raise the possibility that long-term solar irradiance variations may be limited to 11-year cycles.

Solar forcing estimates based on changes in total solar irradiance are only approximations of the actual forcing. This is because of the wavelength dependence of both the magnitude and the variability of the solar spectrum and of atmospheric absorption that differentially attenuates the spectrum. Forcing estimates based on total solar irradiance assume that energy at all wavelengths reaches the Earth's surface (or at least the troposphere) and that radiation at all wavelengths changes by the same amount. However the solar spectrum has less flux, but varies more, at shorter wavelengths. Furthermore, the Earth's stratosphere absorbs solar irradiance at wavelengths less than 310 nm. The energy in the solar spectrum at wavelengths from 200 to 300 nm (15.3 W m^{-2}) changes by about 1 percent during the 11-year cycle and accounts for 13 percent of the corresponding total irradiance cycle (Lean et al., 1997). Variations in solar ultraviolet irradiance alter the production and destruction of ozone, thereby influencing stratospheric temperature, dynamics, and chemistry. The subsequent coupling of the stratosphere with the troposphere (via radiative and dynamical pathways) is considered to produce indirect climate forcing by solar irradiance.

Knowledge of variations in solar spectral irradiance is much poorer than for total solar irradiance. Observations have been made primarily in the ultraviolet spectrum, for only one decade, at wavelengths less than 400 nm. Estimates of variations in the visible and infrared regions have thus far relied on models of the wavelength dependence of the competing sunspot and faculae influences. Only the recently launched SORCE spacecraft has the capability to measure the solar spectral irradiance with the needed long-term precision. Preliminary data already raise questions about the modeled infrared spectrum variability (Fontenla et al., 2003). Whereas current understanding is that faculae are dark in the IR spectrum, SORCE observes increased IR irradiance when faculae are present on the Sun.

Ionization and Production of Cloud Condensation Nuclei

Galactic cosmic rays have one billionth of the total solar irradiance energy, but can reach the troposphere where they produce ions that may serve as nuclei for cloud condensation, with subsequent climatic impacts.

Because the heliosphere influences their transport, cosmic rays exhibit fluctuations that mirror solar activity. When solar activity is high, the more complex magnetic configuration of the heliosphere and the solar wind that flows through it from the Sun to the Earth reduce the cosmic ray flux. There is a close inverse correspondence with solar activity of products of the collision of cosmic ray particles with particles in the Earth's atmosphere (smaller pions, muons), which ground-based neutron monitors have been measuring since the 1950s. The approximate 15 percent modulation of cosmic ray flux by solar activity produces an energy change less than one millionth of the energy change in the 0.1 percent total solar irradiance cycle.

Cosmic rays also interact with air nuclei to produce isotopes such as ^{14}C in tree-rings and ^{10}Be in ice cores. Fluctuations in the ^{14}C and ^{10}Be records (which are superimposed on the larger variations associated with changes in the Earth's magnetic field) are believed to reflect primarily changes in long-term solar activity (Stuiver, 1965; Beer et al., 1990), although climate effects cannot be ruled out (Lal, 1988). During the last 100 years, the ^{10}Be record suggests a 15 percent overall decline in cosmic ray flux.

By altering the population of cloud condensation nuclei and hence microphysical cloud properties (droplet number and concentration), cosmic rays may induce processes analogous to the indirect effect of tropospheric aerosols (Carslaw et al., 2002). Since the plasma produced by cosmic ray ionization in the troposphere is part of an electric circuit that extends from the Earth's surface to the ionosphere, cosmic rays may also affect thunderstorm electrification (Carslaw et al., 2002).

Analysis of cloud cover data reveals decadal variations apparently related to solar-modulated galactic cosmic ray fluxes (Svensmark and Friis-Christensen, 1997), but because solar activity modulates both cosmic ray fluxes and solar irradiance, it is difficult to distinguish which forcing mechanism is responsible for such empirical evidence. The evidence can readily be reinterpreted as association of solar irradiance and cloud cover (Udelhofen and Cess, 2001; Kristjánsson et al., 2002). Using 16.5 years of cloud cover data from the International Satellite Cloud Climatology Project (ISCCP), Kristjánsson et al. (2002) found that low cloud cover correlates better and more consistently with total solar irradiance than with galactic cosmic rays. The data suggest that solar irradiance variations are amplified by interactions with sea surface temperature, which in turn interacts with low cloud cover. In another study, Udelhofen and Cess (2001) found a high coherence between cloud cover inferred from ground-based observations and solar variability over the United States from 1900 to 1987 (but of opposite phase to that found in the ISCCP low clouds). Using cloud coverage simulated by a climate model, they found cloud cover variations in phase with solar variability but not with the galactic cosmic ray flux. They suggest that the cloud variability is affected by a modulation of the atmospheric circulation resulting from variations in the solar-UV-ozone-induced heating of the atmosphere.

Orbital Variation

Earth's distance from the Sun does not remain constant at 1 AU. Rather, the eccentricity of Earth's orbit (currently 0.0167) and the tilt of its axis relative to the orbital plane result in continual changes in the amount and distribution of solar electromagnetic radiation that the Earth receives. In modern times this variation is ±3.5 percent during the year, with maximum energy and minimum distance in January.

Indirect Effects Through the Stratosphere

Of the Sun's mean total radiative output of 1365 W m^{-2}, 15 W m^{-2} (~1 percent) of energy is in the ultraviolet spectrum and does not reach the Earth's surface (e.g., Lean et al., 1997). This energy is deposited in the stratosphere, where it drives ozone formation (and also destruction). Although unavailable for direct forcing of climate, it may induce indirect climate effects as a result of radiative and dynamical coupling of the stratosphere and troposphere. The regional pattern of such indirect climate forcing is likely quite different from the effects of direct surface heating by solar radiation.

The effect of solar cycle UV irradiance changes on stratospheric ozone are now relatively well established as a result of extensive space-based datasets that span more than two solar activity cycles (McCormack et al., 1997). As Figure 2-7 illustrates, the 11-year cycle of ~1 percent peak-to-peak amplitude in middle UV radiation is associated with a 2 to 3 percent modulation of global total atmospheric ozone. The solar UV-induced ozone effects vary with geographic location and altitude, and appear to induce a significant tropopause response (Hood, 2003).

As with tropospheric climate, solar-induced ozone changes occur simultaneously with other natural and anthropogenic effects that must be understood and quantified in order to isolate the solar component (Jackman et al., 1996; Geller and Smyshlyaev, 2002). Most evident is a long-term downward trend in total ozone concentrations associated with increasing concentrations of CFCs. The 11-year solar cycle is superimposed on this trend (Figure 2-7), as are the influences of volcanic aerosols (which warm the stratosphere while cooling the surface), greenhouse gas increases (which cool the stratosphere while warming the surface), and internal variability modes (in particular the quasi-biennial oscillation).

Energetic particles (1 to 100 MeV) produced during eruptive solar events can also produce significant episodic ozone depletion, primarily at higher latitudes (where the particles preferentially enter the Earth's atmosphere) and for relatively short periods (days). Ozone depletion arises from the odd nitrogen chemical destruction cycle that the particles initiate (Jackman et al., 2001). These depletion events, whose frequency and strength vary with solar activity, are superimposed on the more sustained solar UV radiation-induced ozone changes that occur during the 11-year solar cycle.

The extent to which solar UV radiation and energetic particle effects

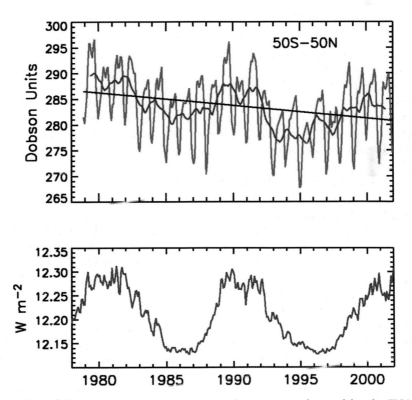

FIGURE 2-7 Top panel: global total atmospheric ozone observed by the TOMS satellite (McCormack et al., 1997). Bottom panel: solar ultraviolet irradiance observed at 200-295 nm (Lean et al., 1997).

have indirect climatic impacts depends on the coupling of the stratosphere with the troposphere. Both radiative and dynamical couplings are surmised (Figure 2-8). Since ozone absorbs electromagnetic radiation in the UV, visible, and IR spectral regions, changes in ozone concentration can affect Earth's radiative balance by altering both incoming solar radiation and outgoing terrestrial radiation. Simulations of this effect (Lacis et al., 1990) show (Figure 2-8) that the net change of surface temperature depends on

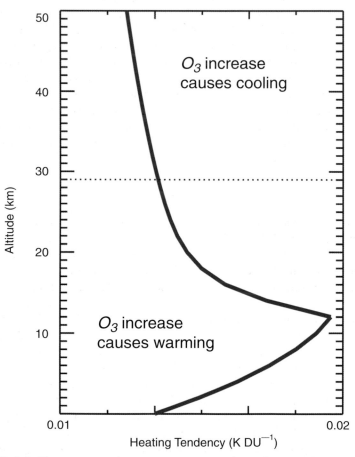

FIGURE 2-8 Change in surface temperature resulting from a change in ozone concentration as a function of altitude. SOURCE: adapted from Lacis et al. (1990).

the altitude of the ozone change; increases below 29 km produce surface warming, and increases above 29 km produce surface cooling. Model simulations suggest that such radiative coupling effects can alter the strength of the Hadley cell circulation, with attendant effects on, for example, Atlantic hurricane flows (Haigh, 2003).

Solar-induced indirect effects on climate may also involve altered modes of variability. Model simulations and analyses of patterns of variability suggest that the Arctic Oscillation (AO), or Northern Annular Mode (NAM), and its subset the North Atlantic Oscillation (NAO) propagate from the stratosphere to the troposphere (Baldwin and Dunkerton, 1999). Radiative forcings that impact the stratosphere could alter this coupling. Contemporary observations suggest that the NAM manifests itself primarily in the North Atlantic sector, as the NAO, during solar cycle minima, and extends more uniformly over all longitudes, as the AO, during solar maxima (Kodera, 2002). The effect of the Sun on the NAM may further depend on the phase of the quasi-biennial oscillation (QBO) in stratospheric equatorial winds (Ruzmaikin and Feynman, 2002). Reduced solar activity in the Maunder Minimum may have produced a negative NAO phase (compared with the current positive phase), based on empirical analysis of historical surface temperature fields and model simulations (Shindell et al., 2001b). Additional evidence that the phase of the QBO changes with the solar cycle (Salby and Callaghan, 2000) underscores the complicated, multifaceted nature of indirect solar effects on climate.

VOLCANIC ERUPTIONS

Emissions from volcanic eruptions have multiple effects on climate as listed in Table 2-3 (Robock, 2002). A number of studies have evaluated the role of volcanic forcing in climate change during the twentieth and earlier centuries (Free and Robock, 1999; Crowley, 2000; Bertrand et al., 2002; Bauer et al., 2003). These studies suggest that volcanic forcing is the dominant source of natural global radiative forcing over the past millennium. The greater prevalence of explosive volcanic activity during both the early and the late twentieth century and the dearth of eruptions over the interval from 1915 to 1960 represents a significant natural radiative forcing of twentieth century climate (e.g., Crowley, 2000). Similarly, the longer-term volcanic radiative forcing has been associated with a significant long-term forced cooling from A.D. 1000 to A.D. 1900 resulting from a general increase in explosive volcanic activity in later centuries (Crowley, 2000; Bertrand et al., 2002; Bauer et al., 2003; Crowley et al., 2003; Hegerl et al., 2003). Some spatially resolved simulations of volcanic forcing indicate a large continental summer cooling but a tendency for a dynamically induced, offsetting winter warming (Stenchikov et al., 2002; Shindell et al.,

TABLE 2-3 Effects of Large Explosive Volcanoes on Weather and Climate

Effect and mechanism	Begins	Duration
Reduction of diurnal cycle Blockage of shortwave and emission of longwave radiation	Immediately	1-4 days
Reduced tropical precipitation Blockage of shortwave radiation, reduced evaporation	1-3 months	3-6 months
Summer cooling of Northern Hemisphere tropics and subtropics Blockage of shortwave radiation	1-3 months	1-2 years
Reduced Sahel precipitation Blockage of shortwave radiation, reduced land temperature, reduced evaporation	1-3 months	1-2 years
Stratospheric warming Stratospheric absorption of shortwave and longwave radiation	1-3 months	1-2 years
Winter warming of Northern Hemisphere continents Stratospheric absorption of shortwave and longwave radiation, dynamics	6-18 months	1 or 2 winters
Global cooling Blockage of shortwave radiation	Immediately	1-3 years
Global cooling from multiple eruptions Blockage of shortwave radiation	Immediately	Up to decades
Ozone depletion, enhanced UV radiation Dilution, heterogeneous chemistry on aerosols	1 day	1-2 years

SOURCE: Robock (2000).

2003). This result contrasts with the response to solar forcing, for which the dynamical and radiative responses appear to reinforce constructively.

Past histories of radiative forcing by explosive volcanic activity are typically constructed from sulfate aerosols contained in annual ice core layers (e.g., Robock, 2000). Spikes of sulfate in the ice core records reflect volcanic injection to the lower stratosphere, where the lifetime is a year or longer, allowing transport to polar regions and eventual deposition after subsidence to the troposphere (Robock and Free, 1995). The longer the residence time of the aerosol in the lower stratosphere, the greater is the associated negative shortwave radiative forcing of the surface through the reflection of radiation back to space. Assumptions must be made regarding the relationship between the sulfate aerosol deposited at the surface and the extent and duration of a significant stratospheric dust veil. These assumptions are highly uncertain and can be only partially tested for a few recent eruptions (e.g., Stenchikov et al., 1998). Greater concentrations of trapped sulfates are typically indicative of larger eruptions, although the proximity of the source region to the ice core may be a complicating factor. Explosive tropical eruptions are more likely to impart a significant global radiative forcing because they provide an opportunity for the aerosol to spread throughout the global lower stratosphere. An eruption is assumed to have occurred in the tropics if its aerosols are recorded in ice cores at both poles.

Other indices of past volcanic activity have also been developed in past work. These include the Volcanic Explosivity Index, or VEI, which is based on qualitative volcanological information and should therefore be used with caution in studies seeking quantitative estimates of climate response (Robock and Free, 1995), and the Dust Veil Index (DVI; see, e.g., Robock, 2000). Some authors argue that the use of climate information in some DVI estimates leads to a potential circularity in using this index to diagnose climate response. Tree-ring reconstructions of continental summer temperature variations have also been used to estimate past volcanic forcing histories (Briffa et al., 1998), although a similar circularity obviously exists if the associated volcanic histories are used to diagnose the climate response to volcanic forcing. Zielinski (2000) and Robock (2000) provide excellent reviews and critiques of various indices of past explosive volcanic activity.

Ice core volcanic radiative forcing estimates have been developed for the past century to the past couple of millennia by numerous researchers (Robock and Free, 1995, 1996; Robertson et al., 1998; Robock, 2000; Crowley, 2000; Ammann et al., 2003; Crowley et al., 2003). The choice of ice cores used to define the volcanic forcing chronology leads to some significant differences among these different estimates. Some of the estimates assume that tropical eruptions dominate the annual global mean radiative forcing (e.g., Free and Robock, 1999; Crowley, 2000), whereas other reconstructions seek to take into account the influence of the latitudinal and seasonal characteristics of the eruptions (Robertson et al., 1998; Ammann et al., 2003).

OCEAN COLOR

Recognized biologically related surface forcings associated with the oceans include ocean color as well as biogenic aerosol emissions (e.g., dimethyl sulfide). Ocean color refers to the radiance backscattered at the air-sea interface. It is determined by water molecules (the blue wavelengths in particular), phytoplankton and detrital particles, and nonbiogenic sediments in coastal waters (Yoder et al., 2001). Bacteria, viruses, colloids, and small bubbles are also possible contributors.

Shell et al. (2003) used a global climate model to assess the role of ocean color in the sea surface temperature and other aspects of the climate system. They found that phytoplankton warm the surface by about 0.05°C on a global average basis. They also found that the large-scale atmospheric circulation is significantly affected by regional alterations of ocean color. These results suggest that the radiative effects of phytoplankton should not be overlooked in studies of climate change.

Frouin and Iacobellis (2002) also determined that absorption of sunlight by phytoplankton must be included in the global radiation budget. They estimated that, compared to pure seawater, the globally and annually averaged outgoing radiative flux is decreased by 0.25 W m^{-2} due to ocean phytoplankton. In coastal and high-latitude regions, the forcing can reach around 1.5 W m^{-2}. They also found that the amount absorbed was species dependent.

TELECONNECTIONS AND RADIATIVE FORCING

Linkages between weather or climate changes occurring in widely separated regions of the globe are referred to as teleconnections. The extent to which regionally concentrated radiative forcing can affect climate via teleconnections is a matter of current research. Determining the importance of regional forcings, such as those from aerosols or land-use change, requires an understanding of the role of teleconnections that can lead forcings in one region to have effects on other regions far away. Teleconnections are most commonly thought of with respect to the transport of energy by atmospheric waves (Tsonis, 2001). For example, regional and global weather patterns have been associated with sea surface temperature anomalies (e.g., Hoerling and Kumar, 2003). Radiative and nonradiative forcing due to regional land-use change can also result in large differences in atmospheric circulation patterns at large distances from the landscape disturbance. For example, land-use change can alter deep cumulonimbus patterns, which affect atmospheric circulation in distant regions (Chase et al., 2000a).

Avissar and Werth (2005) found that deforestation of tropical regions, through teleconnections similar to those produced during El Niño events, has a significant impact on the rainfall of other regions. In particular, they found that the U.S. Midwest is the continental region the most negatively affected by the deforestation of Amazonia and Central Africa during spring and summer, when rainfall decrease could severely damage agricultural productivity in that region. These results are summarized in Figure 2-9. Avissar and Werth (2005) conclude that tropical deforestation considerably

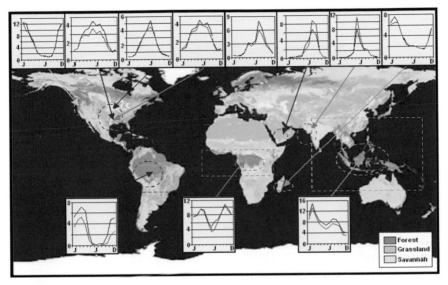

FIGURE 2-9 Annual cycle of precipitation (mm day^{-1}) in continental regions particularly affected by the deforestation of Amazonia (red), Central Africa (green), and Southeast Asia (blue). The blue curves represent the mean monthly precipitation before massive deforestation started in tropical regions (i.e., the "control" case). The red curves indicate the corresponding precipitation following tropical deforestation. The size and location of the color-coded areas corresponding to the deforested regions are at scale. Color-coded ellipses indicate the regions in which tropical forest (in green on the 1-km resolution land-cover map used for the background) was replaced with a mixture of shrubs and grassland. SOURCE: Avissar and Werth (2005).

alters the sensible and latent heat released into the atmosphere and the associated change of pressure distribution modifies the zones of atmospheric convergence and divergence, which shift the typical pattern of the Polar Jet Stream and the precipitation that it engenders.

Radiative forcing by aerosols has also been associated with teleconnected responses in distant locations. For example, a GCM simulation by Chung and Ramanathan (2003) shows that absorbing aerosols over South Asia and the North Indian Ocean can cause subsidence motions over most of the tropics, which would have a drying effect (Figure 2-10).

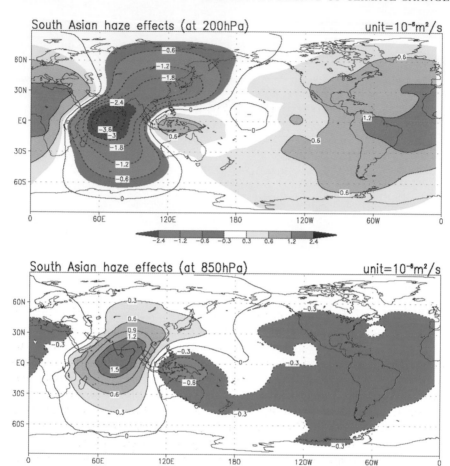

FIGURE 2-10 Velocity potential for the lower troposphere (850 hPa or about 1.5 km) in the lower panel and for the upper troposphere (200 hPa) in the upper panel. The solar heating by absorbing aerosols, mainly due to black carbon, is concentrated over South Asia and the North Indian Ocean (i.e., over the red shaded regions in the lower panel). The red region in the lower panel shows areas of convergence of air or alternately rising motions in response to the solar heating of the lower atmosphere by black carbon. The convergence at the lower levels is followed by divergence (air flowing out of the region) at the upper levels (the blue shaded region in the upper troposphere) over the source region. SOURCE: Adapted from Chung and Ramanathan (2003).

3

Radiative Forcing Over Earth's History

Radiative forcing operates on all timescales over Earth's history (Ruddiman, 2001). This includes (1) geological timescales (tens of millions of years) over which the output of the Sun, concentrations of greenhouse gases, and volcanism vary significantly; (2) timescales of tens to hundreds of thousands of years, over which Earth orbital and astronomical changes appear to dominate the variability of the climate system; (3) timescales of millennia to multimillennia as exemplified by evidence from our present interglacial climate, the Holocene; and (4) modern timescales of approximately the past 1000 years during which natural and anthropogenic influences can be compared. Only for the past few decades are there direct observations (primarily from space) of multiple forcings and the global surface temperature data needed to assess their forcing effects. In this chapter, forcings over each of these timescales are discussed. Tropospheric forcing and response are emphasized because they have been the primary focus of scientific investigation to date.

GEOLOGICAL TIMESCALES

Solar Luminosity

For most of the Sun's 4.5 billion years, the conversion of four hydrogen atoms to one helium nucleus in its core has produced the energy that heats the Sun's surface. The temperature of the Sun's surface determines the amount and spectral shape of the energy that the Sun radiates toward the Earth. According to standard stellar models, as the Sun uses energy, its core density increases and the core shrinks slightly. This causes the core to heat up, which increases the Sun's energy output. Running the standard models backward suggests that at the time of Earth's formation, the Sun was only 70 percent as bright as it is now (Figure 3-1). At the same time however, the Sun was more active, possibly because it rotated three times faster than today, and the rotation helps create the dynamo that drives solar activity, thereby generating stronger magnetic fields (Noyes, 1982). Solar irradiance variations were likely irregular, with large dark spots dominating the surface. Total irradiance therefore likely varied inversely with activity. Although total brightness was lower overall, the ultraviolet (UV) irradiance of the early Sun may have been much higher than today. Solar rotation has slowed because of the loss of mass (and angular momentum) in the solar wind. Solar activity level has decreased because of the resultant decrease in dynamo action, and activity cycles have become more regular.

Today the Sun is middle aged with an anticipated remaining life of about 4 billion years. It rotates once every 27 days and it is brighter, not dimmer, when it is more active and has more spots. This is because mag-

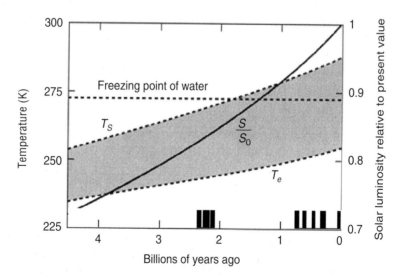

FIGURE 3-1 The Faint Young Sun Paradox. Solid line is solar luminosity relative to present (S/S_0). T_s is Earth's surface temperature and T_e is Earth's effective radiating temperature. Thick vertical bars are glaciations. SOURCE: Modified from Kasting and Catling (2003).

netic fields produce not only dark spots but also bright faculae, whose brightening effects exceed (by a factor of two) the sunspot dimming. As it ages further, the Sun's rotation rate will continue to decrease, but much more slowly because the solar wind is weak. More regular brightness cycles in phase with solar activity and dominated increasingly by bright faculae are expected (Baliunas et al., 1995).

Carbon Dioxide

The weak solar luminosity on the early Earth has posed a long-standing problem whose solution may implicate the greenhouse effect of CO_2 as a long-term temperature regulator. Climate models suggest that for a 30 percent weaker Sun and a modern greenhouse gas concentration, water would have been frozen from 4.5 Ga (billion years ago) until about 2.3 Ga (Sagan and Mullen, 1972), yet geological evidence from well before 2.3 Ga documents a vigorous hydrologic cycle and a diverse marine biota (Sagan and Mullen, 1972; Kasting; 1993; Kump et al., 2000). This contradiction has come to be known as the Faint Young Sun Paradox (Figure 3-1).

The most widely accepted solution to the paradox, based both on models and geological data, is that the early Earth was warmed by a high concentration of greenhouse gases, probably mainly CO_2, perhaps in the range of a few hundred to 1000 times present atmospheric levels (Kump et al., 2000). To explain such high levels of atmospheric CO_2, most attention has focused on the dominant process that draws down atmospheric CO_2 relative to production on geological timescales: the chemical weathering of continental silicate rocks illustrated in the following chemical reaction (Kasting, 1993; Kump et al., 2000; Ruddiman, 2001).

$$H_2O \quad + \quad CO_2$$
Rain   Atmosphere

$$CaSiO_4 \quad + \quad H_2CO_3 \quad \rightarrow \quad CaCO_3 + SiO_2 + H_2O + \tfrac{1}{2} O_2$$
Continental   Carbonic   Shells of
silicates   acid (soils)   marine
                                      organisms

Because silicate weathering rates vary with temperature (by a factor of about two for each 10°C) and area of exposed continental silicate rocks (Ruddiman, 2001), it has been argued that on a cold early Earth with small continental mass (Kump et al., 2000), the negative feedback from severely reduced silicate weathering rates would have led to a build up of volcanic CO_2.

However, if the large CO_2 greenhouse effect remained unchanged as solar luminosity increased, the global temperature of the modern Earth would be too hot for nearly all forms of present-day life (Kump et al., 2000). The suggested solution to this problem is once again a silicate weathering-CO_2 negative feedback. Higher temperatures and larger continents would have increased continental silicate weathering rates, thereby decreasing the CO_2 greenhouse effect and cooling the Earth. Thus, many regard the silicate weathering-CO_2 negative feedback as a thermostat that prevented permanent freezing of the early Earth and, later, prevented permanent temperatures too hot for life.

It should be emphasized, however, that much of what has been said above is based more on models and inference from evidence of warmth on early Earth than on conclusive proxy evidence. In addition, some have argued that methane may have been an important constituent of the early atmosphere (Kasting and Catling, 2003; Rye et al., 1995; Catling et al., 2001; Hessler et al., 2004) and that production of volcanic CO_2 may have been an important source of the early Earth's higher CO_2 greenhouse effect (Ruddiman, 2001). More speculative solutions of the Faint Young Sun Paradox suggest that the early Sun might have been hotter than previously thought (Wuchterl and Klessen, 2001) or that a decreased cosmic ray flux, resulting from an early Sun's stronger solar wind, may have reduced cloud cover and raised global temperatures (Shaviv, 2003).

Geological evidence also suggests that atmospheric CO_2 changed dramatically on timescales of a few to tens of millions of years during the Phanerozoic (Figure 3-2). These changes are on the order of 5 to 10 times present atmospheric level. The record suggests that for at least two-thirds of the last 400 My (million years), levels of atmospheric CO_2 were much higher than at present. It appears that these oscillations in atmospheric CO_2 were linked to recurring changes from greenhouse to icehouse climate states.

The cause of the large atmospheric CO_2 changes during the last 400 My is hotly debated. One view, known as the Berner-Lasaga-Garrels (BLAG) hypothesis, proposes that atmospheric CO_2 changed in response to changes in seafloor spreading rates (Berner et al., 1983). Higher spreading rates increase volcanic activity at both divergent and convergent plate boundaries, thereby increasing the rate of release of volcanic CO_2 from the large rock reservoir of carbon. Plate motion reconstructions for the last 100 My (the limit for which this can be done) suggest that at about 100 Ma (million years ago), spreading rates were 50 percent faster than today; however, during the last 15 My, CO_2 levels fell at the same time that spreading rates increased, calling into question any simple relation between spreading rates and CO_2 (Ruddiman, 2001).

A second view is that plate tectonic-driven uplifts of large plateaus, formation of mountain ranges, and amalgamation of supercontinents (which appear to be associated with low relative sea level) cause large

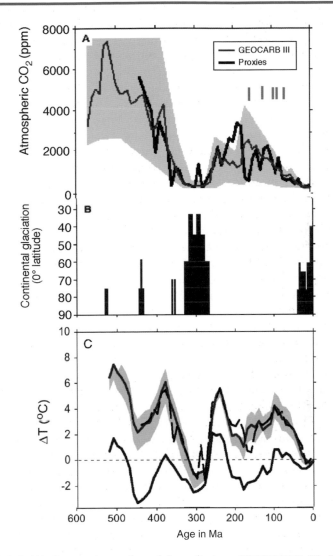

FIGURE 3-2 (A) Comparison of model predictions (GEOCARB III; Berner and Kothavala, 2001) and proxy reconstruction of CO_2 based on a new compilation of 372 published observations (Royer et al., 2004). Shaded area = error of model predictions. Light blue lines between about 160 and 60 Ma are times of brief, geographically limited glaciations in high latitudes (Royer et al., 2004). (B) Latitudinal distribution of major Phanerozoic ice sheets (Crowley, 1998). (C) Comparison of temperature variations from Royer et al. (2004) with those of Shaviv and Veizer (2003). The discrepancy with the Royer et al. (2004) observations suggest that the model of Shaviv and Veizer is incorrect.

increases in the exposure of areas of new, fresh continental silicate rocks to chemical weathering (Kump et al., 2000; Ruddiman, 2001). This is thought to have led to increased silicate weathering rates in the uplifted areas, which as explained in the preceding section would have drawn down atmospheric CO_2. There is, however, little conclusive geological evidence demonstrating that the large landmasses required by the hypothesis actually existed.

It is important to emphasize that if either of the two hypotheses is correct, atmospheric CO_2 changes during the Phanerozoic can be regarded as direct radiative forcings and not feedbacks. That is, just as the source of the CO_2 forcing in the last few decades lies largely outside the climate system (human influences), the source of the Phanerozoic CO_2 forcing lies in changes in plate tectonics that occurred over tens of millions of years.

Shaviv and Veizer (2003) have suggested that changes in cosmic ray fluxes reaching Earth may occur as the solar system passes through the Milky Way's spiral arms and may account, through the influence of cosmic rays on cloud formation, for the major variations in global mean temperature and continental glaciation during the Phanerozoic. A critical examination of the physical and statistical issues underlying this argument, however, reveals that the claimed relationship is likely spurious (Rahmstorf et al., 2004).

Methane

The largest modern reservoirs of methane are methane hydrates buried in sediments of deep lakes and marine continental slopes. The marine reservoir is estimated to be between 500 and 24,000 Gt (billion tons) of carbon, with a best guess assessment of 10,000 Gt (Kvenvolden, 2002). It has been suggested that if enough of the oceanic reservoir were released to the atmosphere by ocean warming or by landslides on the continental slopes, it could produce a brief episode of global warming and a severe perturbation of the carbon cycle. Details of the mechanism have been summarized in Katz et al. (2001).

The most convincing evidence for this type of radiative forcing during the geological past comes from deep and surface ocean records of a remarkably abrupt shift in both oxygen and carbon isotopes dated at about 55 Ma and known as the Paleocene-Eocene Thermal Maximum (PETM). Depletions in $\delta^{18}O$ imply abrupt ocean warmings, perhaps by as much as 3.5 to 4°C in tropical and subtropical latitudes (Tripati and Elderfield, 2004). The exceptionally large depletion in $\delta^{13}C$ of 2 to 4‰ (parts per thousand) is taken to be consistent with a release and subsequent oxidation of between 1500 and 2000 Gt of methane hydrate carbon with a $\delta^{13}C$ signature of about −60‰, a typical value for biogenic methane carbon. Model simulations taking into account both the atmospheric chemistry and the radiative forcing associated with such an increase in methane concentrations successfully reproduce the estimated warming (Schmidt and Shindell, 2003).

If the methane release was triggered by plate-tectonic changes, including possible influences on landslides on continental slopes, as an increasing amount of evidence suggests (Katz et al., 2001), then, as argued for the Phanerozoic CO_2, methane was a radiative forcing agent during the geological past that was produced by a mechanism operating outside the climate system. A recent analysis, however, has revised the amount of marine methane carbon downward to between 500 and 2500 Gt (Milkov, 2004). Kent et al. (2003) have argued that if the revised estimate is correct and applies to the time of the Paleocene-Eocene transition, then ocean methane carbon released by any mechanism probably would have been insufficient to cause a recognizable perturbation of climate and the carbon cycle. They propose instead that a large comet impact might be the more likely explanation for the PETM.

Volcanic Aerosols from Flood Volcanism

Although aerosols produced by strong eruptions of individual volcanoes, such as those along subducting plate boundaries, alter climate for only a year or two at most, aerosols from another category of volcanism, perhaps related to plate tectonics, may force much longer changes in Earth's climate. This type of volcanism, known as flood volcanism, results from deep heating and melting of the mantle at "hot spots." On Earth's surface above the hot spots, the mechanism causes repeated eruptions of basaltic lava that persist for as long as 1 or 2 My. The eruptions produce enormous basaltic plateaus, known as large igneous provinces (LIPs), that range from 7.5×10^5 to 55×10^6 km^3 in volume.

According to Thordarson et al. (2003) and Self et al. (1997), the volatile mass released by flood volcanism is enormous—on the order of 10,000 Mt (megatons) of SO_2 per 1000 km^3 of magma erupted. Amounts of H_2O and CO_2 are likely on the same order as SO_2 emissions. Judging from the volume of individual continental LIPs, the volcanism that produced them released 10,000 Mt of SO_2 into the atmosphere, which is equivalent to 1000 Mt per year for a 10-year-long event. This is a huge amount compared to the 20 Mt released during the eruption of Mt. Pinatubo. By analogy with the Laki flood eruption in Iceland in 1783 and 1784, about 70 percent of the volatiles released at a flood vent are lofted to upper tropospheric-lower stratospheric heights. Thus, the large flood basalt volumes of LIPs over time, particularly on continents, may have caused widespread climate perturbations if the eruption recurrence intervals were shorter than the recovery time of the environment between eruptions. Sparse dating of flows within LIPs precludes any firm estimate of eruption frequency. Even so, claims have been made that a correlation may exist between the timing of major LIP growth and major mass extinctions during the Phanerozoic (Haggerty, 1996; Morgan et al., 2004).

GLACIAL-INTERGLACIAL TIMESCALES

The magnitude and distribution of solar radiation, or insolation, received by the Earth's surface varies due to change in Earth's location and orientation relative to the Sun. Over periods of 100,000 and 400,000 years, the Earth's orbit around the Sun varies from nearly circular (eccentricity = 0.00) to slightly elliptical (eccentricity = 0.06). Received total radiative energy changes by about 0.1 percent as a result of the altered distance of the Earth from the Sun. The tilt, or obliquity, of Earth's orbit, which is currently approximately 23.5° from an axis perpendicular to the plane of orbit, is primarily responsible for the existence of seasons. Variations in the obliquity from 22.1° to 24.5° alter the seasonal distribution of radiation on the Earth's surface with a period near 41,000 years. Polar regions receive greater insolation when the tilt is largest. Precession of the Earth's orbit, which occurs with a roughly 22,000-year periodicity, further modulates seasonality, influencing the relative timing of Earth's closest approach to the Sun (perihelion) relative to the timing of seasons. Currently, perihelion coincides approximately with the Northern Hemisphere winter solstice (favoring decreased seasonal changes in response to seasonal changes in insolation), but the reverse was true 12,000 years ago, at the beginning of the Holocene period.

Periods of approximately 22,000, 41,000, and 100,000 years are prevalent in paleoclimate records and are generally considered to relate, at least in part, to orbital forcing. However, interpreting the dominant 100,000-year cycle in this way is problematic because insolation changes of the order 0.1 percent are too small to have produced the extensive glaciation and cooling of the glacial cycles (Raymo, 1998) and are insufficient to generate the observed variations in model simulations (Kukla and Gavin, 2004). The weak eccentricity forcing must be amplified by the climate system, but the amplifying mechanisms are not well understood. Broecker (1994) noted that glacial terminations are abrupt, in marked contrast to the gradual (sinusoidal) changes in orbital variations. Wunsch (2004) has argued from statistical analyses of climate records that orbital forcing of the 100,000-year glacial cycles accounts for only 20 percent of the variance and is likely indistinguishable from chance. He concluded that broadband stochastic processes are probably the dominant control on glacial cycles.

HOLOCENE

The pre-modern Holocene spans the present interglacial from its onset 11,500 years ago to the time when a reasonable amount of observational data became available, about 1000 years ago. The pre-modern Holocene is regarded as probably the best source of information for understanding how natural radiative forcing agents have changed on timescales of millennia to multimillennia within a modern-like interglacial climate. New evidence suggests that at least regionally the entire ~10,000-year interglacial period contains relatively large oscillations of millennial to multimillennial duration (Hodell et al., 1991; Bianchi and McCave; 1999; deMenocal et al., 2000; Bond et al., 2001; Haug et al., 2001; Thompson et al., 2002; Friddell et al., 2003; Hu et al., 2003; Poore et al., 2003; Niggemann et al., 2003; Magny and Bégeot, 2004).

Orbital-Forced Solar Insolation

Solar insolation changed during the course of the pre-modern Holocene by about 10 to 20 W m^{-2}, in opposite phase for winter and summer. The peak high-latitude summer insolation between about 7,000 and 11,000 years ago likely favored warmer high-latitude summers (sometimes referred to as the Holocene optimum), but cooler high latitude winters and slightly cooler tropical summers, with any net hemispheric or global-scale changes representing a subtle competition between these seasonally and spatially heterogeneous changes (Hewitt, 1998; Kitoh and Murakami, 2002; Liu et al., 2003) and seasonally specific (e.g., vegetation-albedo) feedbacks (e.g., Ganopolski et al., 1998). Recent modeling studies suggest that mid-Holocene global mean surface temperatures may actually have been cooler than those of the mid-twentieth century, even though extratropical summers were likely somewhat warmer (Kitoh and Murakami, 2002).

Extratropical summer temperatures appear to have cooled (Figure 3-3) over the subsequent four millennia (e.g., Matthes, 1939; Porter and Denton, 1967). This period is sometimes referred to as the "Neoglacial" because it was punctuated with periods of glacial advance and retreat of extratropical and tropical mountain glaciers (Grove, 1988). It is reminiscent of, although far more modest than, a full glacial period of the Pleistocene epoch. In many cases, glacial advances appear to have culminated in extensive valley glaciers during the so-called Little Ice Age between the seventeenth and nineteenth centuries.

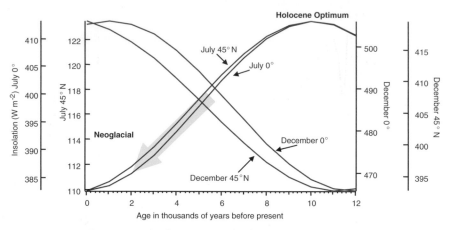

FIGURE 3-3 Changes in insolation over the last 12,000 years at the equator and 45°N for July and December (data from Laskar, 1990). Holocene optimum (warm climate) and Neoglacial (cold climate with increasing numbers of advancing valley glaciers in direction of the large arrow) are climatic events thought to have been associated with the summer insolation changes.

Greenhouse Gases

Prominent multimillennial to millennial time-scale changes in the greenhouse gases CO_2 and methane occurred during the pre-modern Holocene, as documented by measurements from Antarctic and Greenland ice cores (Figure 3-4). Blunier et al. (1995) suggested that changes in methane reflected changes in the hydrological cycle at low latitudes. This interpretation appears to be supported by evidence that the minimum in methane coincides with the time at which many tropical lakes dried up. The subsequent increase in methane is thought to reflect an increasing contribution from northern wetlands as these areas recovered from inhibited growth during earlier, colder temperatures. The large, abrupt decreases in methane in the early Holocene coincide with abrupt coolings in at least the North Atlantic and western European regions. It has been argued recently that interpolar methane gradient data from ice cores is evidence of an abrupt switching on of a major Northern Hemisphere methane source, probably in Siberia, in the early Holocene between about 9,000 and 11,500 years ago (Smith et al., 2004).

Indermühle et al. (1999) interpreted the changes in CO_2 together with $\delta^{13}C$ as evidence of changes in terrestrial biomass and sea surface temperatures. The decline in CO_2 between about 7,000 and 11,000 years ago is

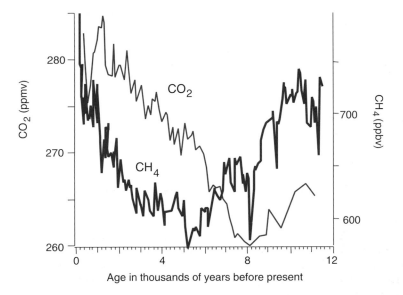

FIGURE 3-4 Changes in the greenhouse gases CO_2 and methane over the last 12,000 years. Methane (CH_4) data from GRIP ice core, Summit Greenland (Blunier et al., 1995), and CO_2 from Taylor Dome ice core, Antarctica (Indermühle et al., 1999).

thought to have occurred as the terrestrial biomass began to increase after the end of the glaciation. This is consistent with the view of Smith et al. (2004) that growth of the large methane source in the early Holocene drew down atmospheric CO_2. The subsequent overall increase in CO_2 to the present is thought to reflect a change to colder and drier conditions in tropical and subtropical regions associated with the Neoglacial trend described above. In this scenario, from about 7,000 years ago to the present, the growing peatlands of the northern latitudes are a source of methane, while the decrease in tropical biomass causes an increase in atmospheric CO_2.

Solar Irradiance

Estimates of solar irradiance variations during the pre-modern Holocene assume that cosmogenic isotope information recorded in tree rings (^{14}C) and in ice cores ([10]Be) provide useful irradiance proxies (Figure 3-5). Yet, whereas the source of irradiance variations are magnetically active regions near the surface of the Sun, cosmogenic isotope variations occur because magnetic fields in the extended solar atmosphere in interplanetary space (the heliosphere) modulate the flux of galactic cosmic rays that reach Earth's atmosphere (see Bard et al., 2000; Crowley, 2000; Webber and Higbie, 2003). Thus, the exact relation between the two is far from clear.

Even so, several recent studies document a relatively close connection at millennial timescales between regional climate proxies and nuclide variations (colored heavy lines in the smoothed records in Figure 3-5). Correlations have been found in a number of Holocene records from regions influenced by the Indian and Asian monsoons, in cave deposits from Europe, in marine sediments from the North Atlantic and the Gulf of Mexico, and in records of precipitation from southwestern Alaska (Bond et al.,

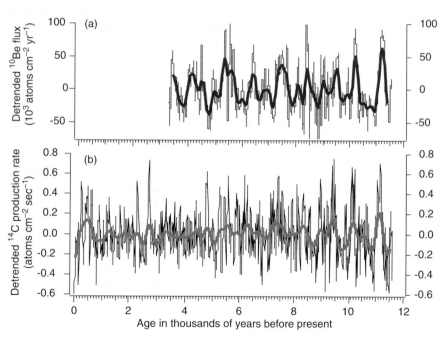

FIGURE 3-5 Changes in the cosmogenic nuclides (a) ^{10}Be and (b) ^{14}C over the last 12,000 years. These changes are taken as proxies of solar activity. Up-pointing peaks indicate reduced activity. ^{10}Be from GRIP/GISP ice cores and ^{14}C from tree-ring measurements. Light black lines are the detrended raw records; heavy colored line represents the same data subject to a binomial smoothing to bring out millennial variability. SOURCE: Adapted from Bond et al. (2001).

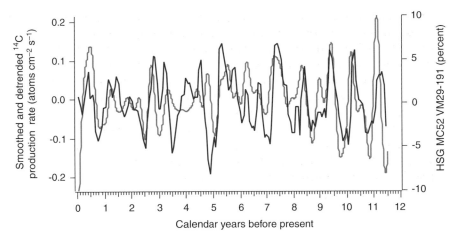

FIGURE 3-6 One example from the eastern North Atlantic of the climate-nuclide connection. Blue lines are smoothed ^{14}C data from Figure 3-5(b); red line is the drift ice record from North Atlantic based on hematite-stained grains (HSG). Up-pointing peaks are relatively colder, possibly by about 1°C, compared to the mean. SOURCE: Adapted from Bond et al. (2001).

2001; Hong et al., 2003; Hu et al., 2003; Niggemann et al., 2003; Poore et al., 2003).

It appears from such correlations that the solar-driven nuclide variations impact climate, at least on a regional scale, enough to leave an imprint in the proxy data (Figure 3-6). These results suggest either that the nuclide-solar irradiance connection is more direct and robust than models suggest, that there are amplified responses of the climate to solar irradiance variations, or both. Potentially, such amplifications include solar ultraviolet impacts on stratospheric ozone and associated tropospheric dynamical responses (e.g., Haigh, 2003; Labitzke and Matthes, 2003; Shindell et al., 2001a, 2003), cosmic ray influences on cloud formation (Carslaw et al., 2002), or changes in North Atlantic meridional overturning (Bond et al., 2001).

LAST 1000 YEARS

Recent theoretical modeling studies have evaluated the role of natural and anthropogenic radiative forcing on climate changes over the past one or more centuries. Detailed attribution studies have focused on the apparent roles of anthropogenic (greenhouse gas and sulfate aerosol) and natural radiative forcings over the spatially data-rich period of the past 100-150 years (Santer et al., 1995, 1996; Tett et al., 1996, 1999; Folland et al., 1998; Hegerl et al., 1997, 2000; Stott et al., 2001). These studies generally find that a combination of natural and anthropogenic forcing is necessary to reproduce early twentieth century changes, while anthropogenic forcing dominates the warming of the latter twentieth century (e.g., Hegerl et al., 2000). However, studies using the instrumental record are limited to a relatively brief (roughly one-century) interval, during which it is difficult to cleanly separate the responses to multiple anthropogenic and natural radiative forcing (Stott et al., 2001).

Longer-term climate model studies have focused on coarser (e.g., hemispheric mean) changes over the past few centuries to millennia (Free and Robock, 1999; Rind et al., 1999; Crowley, 2000; Waple et al., 2002; Shindell et al., 2001b, 2003; Bertrand et al., 2002; Bauer et al., 2003; Gerber et al., 2003; Hegerl et al. 2003; Gonzalez-Rouco et al., 2003) using radiative forcing histories such as those shown in Figure 3-7. The forcing histories include nineteenth and twentieth century anthropogenic radiative forcing (greenhouse gas and sulfate aerosol forcing and, in some cases, land-use changes), and longer-term estimates of natural (volcanic and solar)

FIGURE 3-7 Estimates of natural and anthropogenic radiative forcings over the last couple of millennia used by climate models: (a) forcings used by Crowley et al. (2003), (b) solar and volcanic forcings used by Ammann et al. (2003), and (c) solar and volcanic forcings used by Bertrand et al. (2002). All forcings are expressed in watts per square meter and represent global averages (a and c) and averages for the Northern Hemisphere (b). For panels (b) and (c) the greenhouse gas and sulfate aerosol forcing will be similar to that used in (a). All solar forcing series are expressed as anomalies from the mean value of 1365.6 W m^{-2} (Lean et al., 1995). Details of the extension of the solar series before visual-based observations began in the early seventeenth century are given in Bard et al. (2000) and Crowley (2000). Over this period, the solar forcing in (a) is slightly smaller than the other two because it applies the background trend not to the Maunder Minimum period but to the ^{10}Be estimates for the earlier Spörer Minimum. Volcanic forcing is converted to watts per square meter by multiplying the aerosol optical depth estimates made from ice cores by −21 (Hansen et al., 2002). Volcanic forcing dips below −7 W m^{-2} in either 1258 (panels a and b) or 1259 (panel c) to −9.1, −11.9, and −8.3 W m^{-2}, respectively, in the three panels. SOURCE: Jones and Mann (2004).

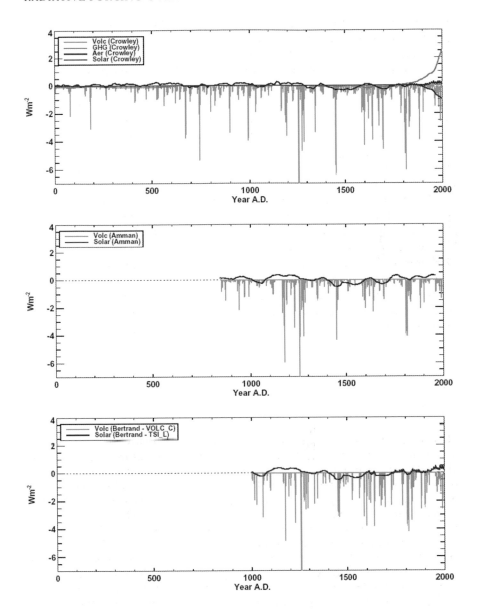

radiative forcing histories over the past 1000 years. These longer-term estimates are based on proxy sources and carry with them certain caveats as discussed in Chapter 2. Figure 3-8 compares the results of simulations of externally forced changes in Northern Hemisphere temperature averages over the past one to two millennia to a reconstruction of Northern Hemisphere mean temperatures from proxy data (Mann and Jones, 2003). These are generally consistent (Figure 3-8) with the reconstruction roughly at the center of the spread of the various model estimates. As the uncertainties in model estimates and climate reconstructions are entirely independent, the level of agreement between the two suggests a significant degree of reliability in the common features between them such as the moderate temperatures from A.D. 1000 to A.D. 1300, the colder conditions from A.D. 1400 to A.D. 1900, and the late twentieth century warmth.

The simulations suggest explosive volcanism is the primary source of changes in natural radiative forcing in past centuries, while anthropogenic forcing increasingly dominates hemispheric mean temperature trends during the nineteenth and twentieth centuries (Hegerl et al., 2003). Solar variability appears to play a significant, although somewhat lesser role, over the same time period (Crowley, 2000; Bertrand et al., 2002; Bauer et al., 2003; Gerber et al., 2003). The combined influence of volcanic and solar forcing appears to provide an explanation of the relatively cool hemisphere mean temperatures from A.D. 1400 to A.D. 1900. Shindell et al. (2003) have argued from model results that regionally—for example, in the North Atlantic and in Western Europe—the climate response to change in solar irradiance may have been more important than volcanism.

During the nineteenth and twentieth centuries, human land-use changes appear to have played a potentially significant role in the large-scale radiative forcing of climate. Bauer et al. (2003) used a climate model to examine the biophysical forcing from deforestation, including increased surface albedo as well as reductions in evapotranspiration and surface roughness. Their simulation can reproduce the actual Northern Hemisphere mean temperatures during the nineteenth and early twentieth centuries, while simulations without this forcing (Crowley, 2000; Bertrand et al., 2002; Gerber et al., 2003) are too warm. Simulations that do not include land-use changes may exhibit an artificially cold pre-nineteenth century mean temperature relative to empirical estimates when, as in Figure 3-8, the model simulation results have been aligned vertically to have the same mean as the instrumental temperature record during the late nineteenth and twentieth centuries. The issue of the role of late twentieth century land-use changes in surface temperature measurements is currently being debated in the scientific literature (e.g., Kalnay and Cai, 2003; Marshall et al., 2004a; Trenberth, 2004; Vose et al., 2004) and is worthy of further investigation.

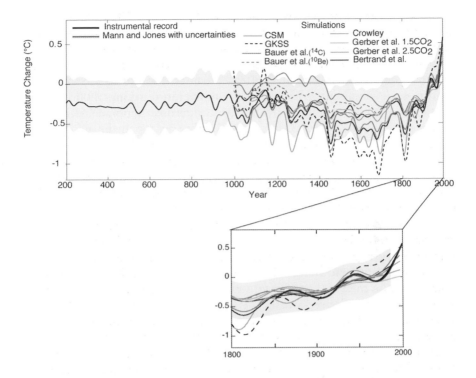

FIGURE 3-8 Model-based estimates of Northern Hemisphere temperature variations over the past two millennia. Shown are 40-year smoothed series. The simulations are based on varying radiative forcing histories, employing a hierarchy of models including one-dimensional energy-based models (Crowley, 2000), two-dimensional reduced complexity models (Bertrand et al., 2002; Bauer et al., 2003; Gerber et al., 2003), and full three-dimensional atmosphere-ocean general circulation (GKSS—Gonzalez-Rouco et al., 2003; CSM—Ammann et al., submitted). Shown for comparison are the instrumental Northern Hemisphere record for 1856-2003 and the proxy-based estimate of Mann and Jones (2003) extended through 1995 (see Jones and Mann, 2004) with its 95 percent confidence interval. Models have been aligned vertically to have the same mean over the common 1856-1980 period as the instrumental series (which is assigned zero mean during the 1961-1990 reference period). The inset provides an expanded view of changes over the past two centuries. SOURCE: Jones and Mann (2004).

Comparisons between long-term model simulations and empirical reconstructions can provide some potential insight into the sensitivity of the climate system to radiative forcing (e.g., Crowley and Kim, 1999). Preliminary climate system modeling, with an emphasis on the carbon cycle of the past millennium (Gerber et al., 2003), indicates that the larger-amplitude century-scale variability evident in some temperature reconstructions is inconsistent with constraints provided by comparison of modeled and observed pre-anthropogenic CO_2 variations. This reinforces the evidence for relatively modest (less than 1°C) variations prior to the twentieth century and for a moderate equilibrium climate sensitivity of roughly 2-3°C for a doubling of CO_2 concentrations (e.g., Cubasch et al., 2001).

Volcanic and solar radiative forcing of changes in the El Niño/Southern Oscillation (ENSO) in past centuries (Ruzmaikin, 1999; Adams et al., 2003; Mann et al., 2005) may explain preliminary empirical evidence for a prevalence of La Niña-like conditions during the eleventh to fourteenth centuries, and El Niño-like conditions during the seventeenth century (Cobb et al., 2003). Such findings, along with anthropogenic land-use change, further emphasize the potential spatial complexity of the climate in response to past changes in radiative forcing.

LAST 25 YEARS

The last 25 years feature unprecedented data documenting simultaneous variations in radiative forcings, climate feedbacks, and climate itself. Many of the more recent datasets have been acquired using space-based instruments, which achieve essentially continuous, global coverage compared to ground-based observations. Space-based observations are available for solar irradiance; volcanic aerosols; concentrations of ozone, CO_2, other greenhouse gases, and CFCs; cloud cover and cloud properties; water vapor; land features, including snow cover, ice, and albedo; temperature of the ocean, land surface, and atmosphere; and other quantities relevant to radiative forcing. Indices are routinely produced, including for ENSO and the North Atlantic Oscillation, based on the extensive datasets and new analysis procedures that extract variability modes. Many of the original databases have been reprocessed, recognizing the need for improved algorithms to remove instrumental drifts so as to better quantify actual change. Examples include the National Centers for Environmental Prediction reanalysis of atmospheric variables, the ISCCP cloud data, and the ground-based Dobson ozone network.

Furthermore, the epoch of the past 25 years is sufficiently long that a range of natural radiative forcing strengths and internal variability modes is sampled concurrently with known anthropogenic forcings. This period includes notable volcanic episodes (i.e., El Chichon, Mt. Pinatubo), two complete solar activity cycles, major ENSO events, land-use changes, and significant increases in greenhouse gases, chlorofluorocarbons, and anthropogenic aerosols. Understanding climate forcings and effects in the last 25 years is a key requirement for securing reliable predictions of future climate change based on forcing scenario studies. Mt. Pinatubo, providing an estimated -3 W m^{-2} global mean surface radiative forcing, is a particularly important test case for examining model-based predictions of response to radiative forcing. The observed surface cooling of roughly 0.4°C has been shown to be consistent with model-estimated responses (Hansen et al., 1992, 2002). Model experiments imposing the inferred vertical radiative forcing profile of Mt. Pinatubo have closely reproduced (Kirchner et al., 1999) the expected seasonal pattern of summer continental cooling and winter warming.

Datasets spanning the past 25 years facilitate a comparison of empirical analysis and model simulations of radiative forcings and their effects on climate. Model experiments employing a combination of anthropogenic and volcanic radiative forcing best match the vertical pattern of temperature changes (Santer et al., 2000) and tropopause height changes (Santer et al., 2003b) over the past couple of decades. A multiple regression analysis of the ENSO index (defined by tropical Pacific sea surface temperatures), volcanic aerosols (according to stratospheric optical depth), solar irradiance (from direct space-based observations), and a linear trend has been argued to reproduce a significant fraction of variability in estimated global lower tropospheric temperatures (Douglass and Clader, 2002). In the latter study, the linear trend was attributed to anthropogenic forcing (a combination of greenhouse gas warming and tropospheric aerosol cooling), while a cooling of 0.5°C was inferred for the Pinatubo eruption and 0.1°C cooling for the solar cycle decrease (forcing of 0.2 W m^{-2}). These latter conclusions must however be treated with caution because other studies using optimal detection approaches indicate that it is difficult to statistically separate the responses to more than two or three distinct natural and anthropogenic forcings even with a century of data (Stott et al., 2001). Moreover, certain indicators used in the study (e.g., ENSO indices) are not physically or statistically independent of the radiative forcings themselves (e.g., Cane et al., 1997; Collins, 2000; Adams et al., 2003; Mann et al., 2005).

Another study of climate change over the past few decades (Hansen et al., 2002) used a general circulation model to estimate forced changes in both surface temperature estimates and the vertical structure of temperature change (the latter as diagnosed from different channels of the MSU satellite observations). The forcings included well-mixed greenhouse gases, stratospheric (volcanic) aerosols, solar irradiance, ozone, stratospheric water vapor, and tropospheric aerosols. The authors found that observed global temperature change during the past 50 years is primarily a response to radiative forcings. They also found that the specification of observed sea surface temperature changes improves the reproduction of the vertical structure of temperature changes. The latter finding is consistent with other studies indicating that SSTs contain additional information (related perhaps to dynamical changes in the climate associated, for example, with ENSO or the annular modes) with regard to the patterns of response to forcing not necessarily produced by the atmospheric response to radiative forcing changes alone (Folland et al., 1998; Sexton et al., 2003).

During the past 25 years, the stratosphere has witnessed significant climate change (Ramaswamy et al., 2001; Shine et al., 2003; Ramaswamy and Schwarzkopf, 2002; Schwarzkopf and Ramaswamy, 2002). Indeed, the temperature at 1 mb has dropped significantly over the past two decades. Models are able to explain the cooling of the lower stratosphere in terms of stratospheric ozone loss; upper stratosphere cooling is due to a combination of ozone change and greenhouse gas increases.

4

Rethinking the Global Radiative Forcing Concept

The current global mean top-of-the-atmosphere (TOA) radiative forcing concept with adjusted stratospheric temperatures has both strengths and limitations. The concept has been used extensively in the climate research literature over the past decades and has also become a standard tool for policy analysis endorsed by the Intergovernmental Panel on Climate Change (IPCC). The concept should be retained as a standard metric in future climate research and policy. However, it also has significant limitations that have been revealed by recent research on forcing agents that are not conventionally considered and by regional studies. Also, it diagnoses only one measure of climate change (equilibrium response of global mean surface temperature). The committee believes that these limitations can be addressed through the introduction of additional forcing metrics. Table 4-1 gives a list of these metrics and summarizes their strengths and limitations. Detailed discussion of each is presented below.

THE CURRENT CONCEPT

Global-annual mean adjusted radiative forcing at the top of the atmosphere is, in general, a reliable metric relating the effects of various climate perturbations to global mean surface temperature change as computed in general circulation models (GCMs). The associated climate sensitivity parameter λ varies by only about 25 percent within a particular GCM[1] for a

---

[1]Estimates of climate sensitivity by different GCMs vary by about a factor of two, ranging from 1.5 to 4.5 K (IPCC, 2001).

TABLE 4-1 Metrics for Climate Forcing

Climate Forcing Metric	Strengths	Limitations
IPCC TAR adjusted tropopause or TOA global average radiative forcing with adjustment of stratospheric temperatures	• Changes in global mean surface temperature are nearly linearly related to global mean TOA radiative forcing for a wide range of forcing agents • Simple and computationally efficient • Enables comparison of different forcing agents • Enables comparison of different models with one another, with benchmarks, and with estimates in the literature • Can be used in simple climate models for policy analysis • Already introduced into the policy dialogue	• Conveys insufficient information about hydrological response • Does not fully characterize the climate impact of light-absorbing aerosols • Does not characterize regional response • Does not accommodate nonlinear response from large perturbations • Does not fully characterize the climate impact of nonradiative forcing, the indirect aerosol effect (other than the first), and the semidirect aerosol effect
Radiative forcing calculated with fixed sea surface temperature (SST; Hansen et al., 2002) or fixed surface temperatures (Shine et al., 2003) and adjusted atmospheric temperatures (in both troposphere and stratosphere)	• Allows calculation of indirect and semidirect effects of aerosols • Incorporates fast atmospheric feedbacks in the simulation of climate forcing and response • Insensitive to the altitude at which forcing is calculated	• Subject to limitations of the standard TOA forcing except for ability to calculate indirect and semidirect effects of aerosols • Not as computationally expedient as the standard radiative forcing calculation • Not as readily comparable across models as the standard radiative forcing calculation
Global mean radiative forcing at the surface	• Provides a characterization of the surface energy budget • If reported with TOA forcing, may provide information on how forcing affects the lapse rate, with implications for precipitation and mixing	• Does not allow characterization of regional structure

TABLE 4-1 Continued

Climate Forcing Metric	Strengths	Limitations
Regional radiative forcing (direct and indirect)	• May provide a better measure of regional climate response than global radiative forcing • Allows characterization of teleconnected response to a regionally isolated forcing	• Further work is needed to quantify links of regional radiative forcing to regional and global climate response
Regional nonradiative forcing (hydrological, land use, biogeochemical)	• Recognizes additional nonradiative climate forcing components • Allows characterization of teleconnected response to a regionally isolated forcing • Nonradiative forcing alters radiative forcing and thus provides a more complete characterization of radiative forcing	• No widely accepted metrics for quantifying regional nonradiative forcing • Further work is needed to quantify links of regional nonradiative forcing to regional and global climate response • Some types of nonradiative forcing are not easily quantified in watts per square meter, thus it is not clear how to compare them to radiative forcing
Ocean heat content	• Can be used to calculate the net radiative imbalance of the Earth • Offers a valuable constraint on the performance of climate models	• Observations may have insufficient frequency and spatial coverage to accurately determine the radiative imbalance at the necessary resolution

NOTE: TAR = IPCC Third Assessment Report (IPCC, 2001).

wide range of changes in well-mixed greenhouse gases, solar irradiance, surface albedo, and nonabsorbing aerosols (IPCC, 2001). By assuming a constant climate sensitivity parameter, forcing can be translated directly into a temperature response. Because calculating radiative forcing is straight-forward, many factors that may influence radiative forcing can be investigated in climate models, simpler versions of those models, and chemical transport models. For example, the effects of different estimates of past forcings can be compared to each other. Likewise, comparisons can be made of multiple possible future forcing scenarios. Furthermore, the radiative forcing concept facilitates comparison of forcing calculations between climate models and with benchmark line-by-line radiative transfer calculations.

Radiative forcing is thus one of the more highly quantified methods of determining how the climate system is forced. In addition, observational records are available for surface temperature (space-based monitoring, in situ monitoring, and proxy data) and the radiation balance at the top of the atmosphere. These data provide an important observational constraint on estimates of radiative forcing and temperature response. Furthermore, numerous model and observation-based estimates of radiative forcing have been reported in the scientific literature over the past decades, providing an important historical reference for future calculations.

The radiative forcing concept has also been used effectively in policy applications. The concept is already entrained in the policy dialogue, particularly through the emphasis given it in the IPCC reports. Policy analysts have input radiative forcing into simple climate models, which are used to examine a wide range of scenarios of past, present, and future climate. Comparison between these simple models and the more complex fully coupled models also helps in interpreting causal mechanisms in the fully coupled models (e.g., Murphy, 1995; Raper et al., 2001).

Although the traditional TOA radiative forcing concept remains very useful, it is limited in several ways. It is inadequate to describe fully the radiative effects of several anthropogenic influences including

• absorbing aerosols, which lead to a positive radiative forcing of the troposphere with little net radiative effect at the top of the atmosphere;
• effects of aerosols on cloud properties (including cloud fraction, cloud microphysical parameters, and precipitation efficiency), which may modify the hydrological cycle without significant radiative impacts;
• perturbations of ozone in the upper troposphere and lower stratosphere, which challenge the manner in which the stratospheric temperature adjustment is done; and
• surface modification due to deforestation, urbanization, and agricultural practices and surface biogeochemical effects.

Land surface modification of heat fluxes and aerosol-induced changes to the precipitation efficiency modify not only the radiative fluxes but also the dynamical (turbulent heat flux) and thermodynamical fluxes (evaporation). These modifications to the climate system fall under the broader umbrella of *climate forcings*, which include radiative and nonradiative fluxes. Broadening the concept of radiative forcing in this way allows consideration of climate variables that may have more direct societal impacts, such as changes in precipitation. Indeed, the traditional radiative forcing concept is inappropriate to predict the sign or the magnitude of the global mean precipitation changes due to both scattering and absorbing aerosols, which affect precipitation differently in summer and winter.

Another limitation of the traditional radiative forcing concept is that it does not adequately characterize the regional response. Regional radiative forcings from atmospheric aerosols, tropospheric ozone, or land-use and land-cover changes can be much larger than global mean values. A regionally concentrated forcing may lead to climate responses in the region, in another region via teleconnections, or globally—or may even have no climate response.

Yet another limitation of the concept is that the assumption of a constant, linear relationship between changes in global mean surface temperature and global mean TOA radiative forcing does not always hold. This linear relationship breaks down for absorbing aerosols, which may have small TOA forcing, but disproportionately larger surface forcing due to absorption of solar radiation (Lohmann and Feichter, 2001; Ramanathan et al., 2001a). This motivated the introduction of the concept of efficacies of different forcing agents (Joshi et al., 2003; Hansen and Nazarenko, 2004). "Efficacy" is defined as the ratio of the climate sensitivity parameter λ_i for a given forcing agent to λ for a doubling of carbon dioxide (CO_2) ($E = \lambda_i / \lambda_{CO_2}$). The efficacy E is then used to define an effective forcing $f_e = f\, E$. Table 4-2 summarizes the forcings, responses, efficacies, and effective forcings of different forcing agents from several models. Efficacies greater than 1, such as for black carbon impacts on snow and ice albedo, correspond to a larger effective forcing than that of $2 \times CO_2$ (Table 4-2). On the other hand, scattering sulfate aerosols are less efficient than greenhouse gases in changing the surface temperature for a given forcing.

Overall, after weighing the strengths and limitations of the traditional radiative forcing concept, the committee finds that its strengths warrant continued use in scientific investigations, climate change assessments, and policy applications. The concept is relatively easy to use, particularly in enabling efficient comparisons between different forcing agents, forcing scenarios, and climate models. Further, it has clear applications within the climate policy community. Nonetheless, the limitations call for broadening the concept to account for nonradiative forcing, spatial and temporal heterogeneity of forcing, and nonlinearities. This chapter presents specific approaches to address these limitations.

GLOBAL MEAN RADIATIVE FORCING WITH ADJUSTED TROPOSPHERIC TEMPERATURES

Hansen et al. (2002) introduced the concept of fixed sea surface tem-

TABLE 4-2 Efficacies and Effective Forcing Calculated for a Variety of Forcings in Several Climate Models

Experiment	Forcing at the Tropopause f (W m^{-2})	Equilibrium Response at the Surface ΔT (K)	Efficacy E	Effective Forcing f_e (W m^{-2})	Reference
CO_2 (based on three GCMs)	1	0.38-1.12	1	0.38-1.12	Joshi et al. (2003)
Changes in snow and ice albedo due to black carbon (two scenarios)	0.16-0.17	0.2-0.24	1.97-2.22	0.32-0.38	Hansen and Nazarenko (2004)
Solar increase based on three GCMs	1	0.31-1.07	0.82-1.01	0.82-1.01	Joshi et al. (2003)
Solar increase	4.2-3.7	2.1-2.85	0.52-0.8	2.2-3.0	Gregory et al. (2004)
Tropospheric ozone	1	0.24-0.94	0.71-0.88	0.71-0.88	Joshi et al. (2003)
Tropospheric ozone	0.49	0.28	0.73	0.49	Mickley et al. (2004)
Sulfate aerosols, direct effect	-0.34	-0.24	0.83	-0.28	E. Roeckner (personal communication, May 2004)
Sulfate aerosols, first indirect effect	-0.89	-0.78	1.01	-0.90	E. Roeckner (personal communication, May 2004)
Well-mixed greenhouse gases from 1860 to 1990	2.12	1.82	1	2.12	Feichter et al. (2004)
All aerosol effects (direct and indirect on water clouds)	-1.4	-0.87	0.72	-1.01	Feichter et al. (2004), Lohmann and Feichter (2001)
Greenhouse gases + direct sulfate aerosol forcing (1900-present) based on four GCMs	—	0.5-0.7	0.56-0.75	—	Boer et al. (2000)
All aerosol effects and greenhouse gases	0.7	0.57	0.94	0.66	Feichter et al. (2004), Lohmann and Feichter (2001)

perature (SST) forcing, which is the change in TOA radiative forcing computed in a global model with fixed sea surface temperatures but letting land and atmospheric temperatures relax to the new equilibrium. This relaxation is relatively rapid (on the order of years); hence the calculation in a GCM is computationally expedient. One resolves in this manner the short-term components of the climate response, such as hydrological perturbations associated with changes in lapse rate. Of particular interest, this approach allows calculation of a meaningful radiative forcing from the indirect or semidirect aerosol effects. Hansen et al. (2002) show that there is good agreement between the stratospheric adjusted radiative forcing and the fixed SST forcing for a range of climate forcing factors (e.g., $2 \times CO_2$, stratospheric aerosols) and that for changes in ozone, more reasonable forcings result from the fixed SST simulations.

Shine et al. (2003) extended the fixed SST approach to what they call "adjusted troposphere and stratosphere forcing." Shine et al. not only fix sea surface temperatures, but also fix land surface temperatures because temperatures over land and ocean are related. Therefore, it is more consistent to fix surface temperatures globally. Using a global climate model they show that the adjusted troposphere and stratosphere radiative forcing is consistent with the stratospheric adjusted forcing for more uniform forcings such as doubling CO_2 and solar constant changes. They also show that for forcings due to absorbing aerosols, their newly defined forcing is more meaningful than the stratospheric adjusted forcing, in that the climate sensitivity parameter is largely independent of how the absorbing aerosols are vertically distributed, unlike the standard stratospheric adjusted approach.

These two studies are important contributions to the debate on radiative forcing, but the approach is subject to most of the limitations associated with the traditional radiative forcing calculation. Also, forcings calculated in this manner are not as easy to compute as conventional radiative forcings, nor are they as comparable among different GCMs because of differences in model dynamics and hydrology.

GLOBAL MEAN RADIATIVE FORCING AT THE SURFACE

The TOA radiative forcing might not be directly related to surface temperature if a forcing agent changes the vertical distribution of heating in the atmosphere. Well-known examples of such cases are the direct radiative forcing of black carbon (BC) and other absorbing aerosols and the changes in latent and sensible heat fluxes due to land-use modifications. For example, BC causes an increase in atmospheric heating, accompanied by a decrease in solar heating of the surface. For average cloudiness, Indian Ocean Experiment (INDOEX) data reveal that the TOA direct forcing when BC is present can be close to zero, while the surface forcing can be on

BOX 4-1
Two Case Studies of Regional Aerosol Forcing

South Asian Pollution Observed over the North Indian Ocean

The figure below shows the anthropogenic aerosol forcing over the North Indian Ocean averaged for January-April during 1996 to 1999 for available INDOEX observations (Ramanathan et al., 2001a). The bar chart shows the direct forcing, the indirect forcing, and the greenhouse forcing at the top of the atmosphere, the surface, and the atmosphere. The sum of the direct and the indirect forcing at the surface is as much as -20 W m^{-2}, which amounts to a reduction of about 10 percent of the absorbed solar radiation at the surface. A correspondingly large positive forcing of 15 W m^{-2} is exerted on the atmosphere. The positive atmospheric forcing is due largely to the soot and dust absorption of solar radiation. The TOA forcing is a small difference (-5 W m^{-2}) between two large competing terms at the surface and the atmosphere. If only the TOA radiative forcing is considered, one would conclude that the direct climate effect of Asian aerosols is near zero.

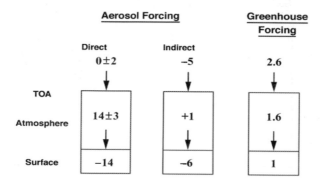

Four-year (1996 to 1999) average of direct and indirect aerosol forcing during the dry season (January to April) observed during the INDOEX campaign. The data are averaged over the North Indian Ocean, from the equator to 25ºN and from 60ºE to 90ºE. SOURCE: Ramanathan et al. (2001a).

the order of -10 to -15 W m^{-2} and the atmospheric forcing on the order of $+10$ to $+15$ W m^{-2} (see Box 4-1). For such aerosols, the TOA forcing is an ineffective, if not erroneous, metric for the impact of aerosol forcing on the surface temperature.

One way to address this limitation of the traditional radiative forcing concept is to calculate the global mean radiative forcing at the surface along with that at the top of the atmosphere. Considering the surface radiative forcing may enable quantification of the effects of aerosols on the surface

Yet observational and modeling studies have shown that these aerosols have led to large regional changes in surface and atmospheric temperatures, the surface energy budget, and rainfall (Ramanathan et al., 2001a; Chung et al., 2002; Menon et al., 2002b).

East Asian Pollution Observed over the Western Pacific Ocean

Similar regional averaged forcing values computed in a model constrained by observations have been obtained for the western Pacific region off East Asia using data from the Aerosol Characterization Experiment in Asia (ACE-Asia) field campaign (Conant et al., 2003). The figure below shows direct surface forcing, atmospheric forcing, and TOA forcing for the region averaged over 10 days of observations. The magnitudes are comparable to those obtained for the Indian Ocean.

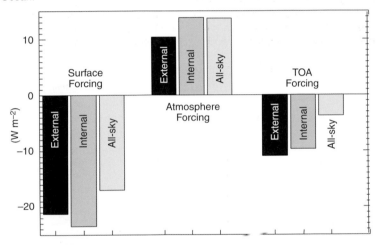

Direct aerosol forcing for April 5 to 15 computed with a model constrained by ACE-Asia observations. The values are averaged over 20ºN to 50ºN and 100ºE to 150ºE. Results are shown from three versions of the model: externally mixed aerosol and clear-sky conditions; internally mixed aerosol and clear-sky conditions; and all-sky (clear and cloudy) conditions. SOURCE: Conant et al. (2003).

energy balance. Together with the TOA radiative forcing, surface radiative forcing also may provide information about the extent to which forcings affect the atmospheric lapse rate, with implications for precipitation and mixing. The net radiative forcing of the atmosphere could be deduced from the difference between TOA and surface forcing. Like the other metrics discussed above, global mean radiative forcing at the surface would not allow characterization of the regional structure of forcing.

Reporting surface radiative forcing along with that at the TOA is important not just for absorbing aerosols but also for the gaseous species. Traditionally, the notion has been that it is enough to give the tropopause forcing for the well-mixed gases in order to obtain an estimate of the surface temperature response. However, to assess climate response beyond surface temperature change (e.g., changes in precipitation, latent heat release from surface, or in the surface heat and moisture balance), it becomes necessary to understand the surface radiative forcing for all forcings. Further, to understand the difference in the zonal-mean hydrologic response between different forcings, it is necessary to look at the surface terms (e.g., Chen and Ramaswamy, 1996).

REGIONAL RADIATIVE FORCING

Forcings with significant spatial variability can have regional magnitudes much greater than their global averages. Aerosols, and to a lesser extent tropospheric ozone, have shorter lifetimes than the well-mixed greenhouse gases, and therefore their concentrations are higher in source regions and downwind (e.g., Charlson et al., 1991; Kiehl and Briegleb, 1993; Mickley et al., 1999). Forcing due to land-use and land-cover changes also has significant spatial heterogeneity, leading to spatial variability in the associated climate response. The traditional global mean radiative forcing provides no information about this regional structure, so many researchers have begun to present estimates of radiative forcing on a regional scale as derived from models or observational campaigns.

A large number of modeling studies have been carried out to characterize the spatial variability in aerosol forcing due to direct, indirect, and semidirect effects (IPCC, 2001). Regional effects of aerosol forcing are large; regional mean values of anthropogenic aerosol radiative forcing can be factors of 5 to 10 higher than the global mean values of 0.5 to 1.5 W m^{-2} (IPCC, 2001). Comparisons with satellite radiation budget data can be used to constrain model results. For example, the calculations of Haywood et al. (1999) showed that the clear-sky outgoing flux at the TOA over oceans yields excellent agreement with Earth Radiation Budget Experiment (ERBE) observations when aerosol species are considered. This is a useful test of the chemical transport model (CTM)-derived concentrations of aerosols and assumptions about their sizes, at least in terms of their collective reflective ability. More recent computations from the National Center for Atmospheric Research and Geophysical Fluid Dynamics Laboratory models bear this out with updated CTM simulations. Soden and Ramaswamy (1998) inferred the existence of spatial aerosol effects in satellite datasets. Observations over source regions and downwind show very large forcings (see Box 4-1). High regional concentrations of scattering aerosols can completely offset the positive forcing due to increases in greenhouse gases. This offset could even be significant on the global scale (e.g., Anderson et al., 2003a).

The degree of spatial heterogeneity can be seen by considering the aerosol optical depth for a number of aerosol species as shown by model results in Figure 4-1. The large optical depths off of the Sahara are due to mineral dust, while the large optical depths in South America and Africa are related to biomass burning. Large aerosol optical depths due to sulfur emissions occur in Northern Hemisphere industrial regions. These optical depths can be used in conjunction with assumptions about aerosol radiative properties to calculate the direct forcing. Results are shown in Figure 4-2 for TOA, surface, and atmospheric radiative forcings. Note the significant difference between the TOA forcing and the effect at the surface due to the absorptive properties of the aerosols. Regional forcing values at the surface can be as large as -20 to -30 W m^{-2}.

The consequences of regional radiative forcing on the climate system for a region, for other regions, and globally must be better understood. Because global forcings can also have regionally specific responses, it is difficult to attribute regional climate changes to a particular regional or global forcing. A further complication is that regional diabatic heating results in nonlinear long-distance communication of convergence and divergence fields, often referred to as teleconnections. For example, Chase et al. (2000a) found that regional land-use change can cause significant climate effects in other regions through teleconnections, even with a near-zero change in global averaged radiative flux. Chen and Ramaswamy (1996) and Ramaswamy and Chen (1997) showed that significant responses in precipitation patterns can arise in the presence of a near-zero global change in radiative forcing. Regional radiative forcing may provide a better measure of regional climate response than global radiative forcing, but further work is needed to quantify the links of regional radiative forcing to regional and global climate response.

REGIONAL NONRADIATIVE FORCING

Some forcings affect the climate system in nonradiative ways, in particular by modifying the hydrological cycle or vegetation dynamics. These nonradiative forcings generally have radiative impacts, but describing them only in terms of this radiative impact does not convey fully their influence on climate variables of societal relevance. For example, aerosol-induced changes in precipitation may have a small net effect on TOA radiative forcing, but could have significant impacts on the amount of rainfall a region receives, with consequences for agriculture, flood control, and municipal water supply. Furthermore, quantifying nonradiative forcings in

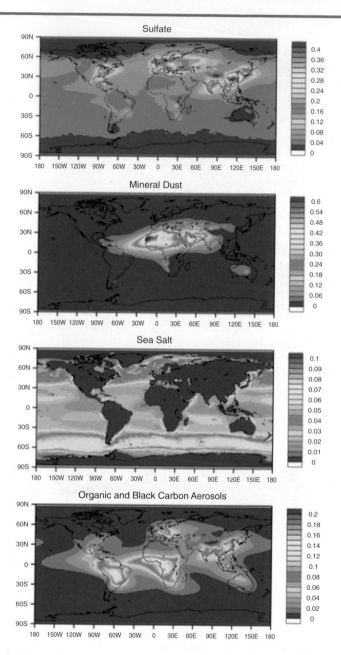

FIGURE 4-1 Annual mean aerosol optical depth predicted by an aerosol chemical transport model due to sulfate, mineral dust, sea salt, and organic and black carbon aerosols. SOURCE: Collins et al. (2002).

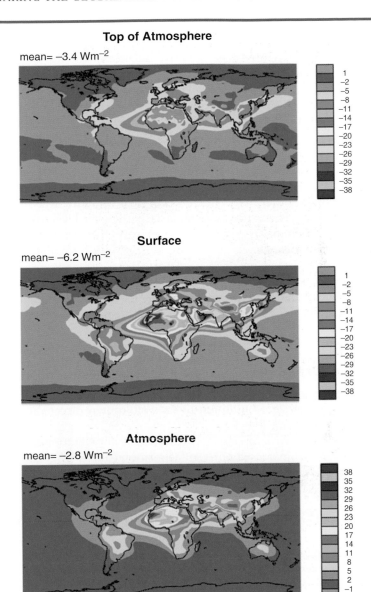

FIGURE 4-2 Annual mean simulated clear-sky radiative forcing in watts per square meter at the top of the atmosphere (net), surface (net), and in the atmosphere calculated from the aerosol optical depths shown in Figure 4.1. SOURCE: Adapted from Collins et al. (2002).

terms of their ultimate radiative effects (i.e., change in TOA or surface radiative fluxes) is not always straightforward.

Aerosols not only affect the radiative balance at the top of the atmosphere but also exert a forcing on the hydrological cycle (e.g., Ramanathan et al., 2001a). An increase in aerosols of similar hygroscopicity leads to an increase in cloud droplet number concentration that reduces the precipitation efficiency for warm clouds. Thus, these clouds will produce fewer drizzle-size drops (second indirect effect). On the other hand, if some of the anthropogenic aerosols act as ice nuclei, supercooled clouds could be converted into ice clouds by the glaciation indirect effect (Lohmann, 2002), resulting in more efficient precipitation formation. Smaller cloud droplets could also exhibit a thermodynamic forcing by protracting freezing in deep convective clouds. These clouds then glaciate in higher levels, which could result in either more or less vigorous precipitation formation depending on the background aerosol levels and atmospheric stability (Khain et al., 2004). Rosenfeld (2000) refers to these kinds of aerosol forcings as *thermodynamic forcing* because the spatial patterns of diabatic heating are changed.

Several nonradiative forcings involve the biological components of the climate system. They can be categorized into three types:

1. *Biophysical forcing* involves changes in the fluxes of trace gases and heat between vegetation, soils, and the atmosphere. For example, in the presence of increased CO_2, plants open their stoma less and are therefore more water efficient (e.g., Sellers et al., 1997). Thus, increased CO_2 impacts the hydrological cycle, in addition to its well-known direct radiative impacts.

2. *Biogeochemical forcing* involves changes in vegetation biomass and soils. For example, increased nitrogen deposition caused by greater anthropogenic emissions of ammonia (NH_3), nitric oxide (NO), and nitrogen dioxide (NO_2) is a biogeochemical forcing of the climate system (Holland et al., 2004). This deposition has altered the functioning of soil, terrestrial vegetation, and aquatic ecosystems worldwide. Galloway et al. (2004) document that human activities increasingly dominate the nitrogen budget at the global scale and that fixed forms of nitrogen are accumulating in most environmental reservoirs. In addition to impacts on ecosystem functioning, which are important in themselves, this forcing modifies physical components of the climate system, such as surface albedo and sensible and latent heat.

3. *Biogeographic forcing* involves alterations in plant species composition. Such changes can occur slowly in response to changes in the weather over time, or suddenly due to fires or other disturbances. For example, greater shrub growth in the high latitudes of the Northern Hemisphere has been observed (McFadden et al., 2001), which could alter the spatial distri-

bution of drifting snow and subsequent melt pattern and timing (Liston et al., 2002).

These forcings are not yet well understood and are the subject of active research (Cox et al, 2000; Betts, 2001; Friedlingstein et al., 2001). They will likely be associated with multiple types of climate responses and are not expected to be additive to the traditionally defined forcings. Complex interactions among these forcings make it difficult to determine their net climate effects (Eastman et al., 2001b; Narisma et al., 2003; Raddatz, 2003). Eastman et al. (2001b), for example, found that with doubled CO_2 the grasslands of the central United States were more water efficient on an individual stoma level (biophysical forcing), but grew more biomass (biogeochemical forcing). The net effect was cooler daytime temperatures during the growing season.

There are no widely accepted metrics for quantifying regional nonradiative forcing. Indeed, because nonradiative forcings affect multiple climate variables, there is no single metric that can be applied to characterize all nonradiative forcings (Marland et al., 2003; Kabat et al., 2004). Nonradiative forcings generally have significant regional variation, making it important that any new metrics be able to characterize the regional structure in forcing and climate response—whether the response occurs in the region, in a distant region through teleconnections, or globally. As is the case for regional radiative forcing, further work is needed to quantify links between regional nonradiative forcing and climate response. Another consideration in devising metrics for nonradiative forcings is enabling direct comparison with radiative forcings, computed in units of watts per square meter. However, not all nonradiative forcings are easily quantified in these units.

A metric that could prove useful for quantifying impacts on the hydrological cycle is changes in surface sensible and latent turbulent heat fluxes. For example, Pielke et al. (2002) proposed the surface regional climate change potential (RCCP), which is calculated by summing and weighting globally the absolute values of changes in the surface sensible and latent turbulent heat fluxes. In their study, land-use change from the natural to the current global landscape produced a global average RCCP of 0.7 W m^{-2} when teleconnection effects were not included, and 8.9-9.5 W m^{-2} when teleconnections were included. Such a scaling of the land surface forcing provides a metric that can be expressed in the same units as radiative forcing. Extending this concept to the global water cycle, Pielke and Chase (2003) quantified landscape forcing in terms of precipitation and moisture flux changes. They found globally averaged differences between the current and the natural landscape of 1.2 mm day^{-1} for precipitation and 0.6 mm day^{-1} for moisture flux. However, such metrics do not provide a complete measure of the integrated effect on the climate system due to the regional concentration of changes in diabatic forcing. Others have considered using more comprehensive model output to quantify the impact of human disturbance to the climate system (Claussen et al., 2002).

OCEAN HEAT CONTENT

The ocean is the largest heat reservoir in the climate system (Levitus et al., 2000, 2001). Thus, the change in ocean heat storage with time can be used to calculate the net radiative imbalance of the Earth (Ellis et al., 1978; Piexoto and Oort, 1992). In essence, the ocean heat content provides a metric for the integral in time of the TOA radiative forcing. Furthermore, it offers a valuable constraint on the performance of climate models (Barnett et al., 2001). It is not yet standard practice to use ocean heat content observations, which are available for the past 50 years, to validate forced climate simulations. This is in part because there are several open research questions regarding the accuracy with which ocean heat content can be calculated and applied.

It is not clear, for example, that the observation systems have sufficient frequency and spatial coverage to accurately determine the radiative imbalance on an annual basis (on the order of 0.1 W m^{-2}) so as to independently confirm the calculation of radiative imbalance from the changes in ocean heat storage. Another issue is whether the spatial and temporal sampling of the ocean heat content accurately captures the regions and depths at which heat changes are occurring. In particular there could be significant heat storage changes deeper in the ocean that are inadequately monitored by the existing ocean network.

Several estimates of the trend in ocean heat content have been made using the ARGO network of ocean floats, satellite observations of ocean altimetry (Levitus et al., 2000, 2001; Willis et al., 2003), and climate models (Barnett et al., 2001; Crowley et al., 2003). Not all of these studies express the ocean heat content changes in terms of average radiative forcing, although it is straightforward to do so. Pielke (2003) found that for the period 1955-1995 the imbalance was about 0.3 W m^{-2}, with half between the surface and 300 m, and the rest between 300 m and 3 km. He also found large temporal variations in the imbalance with a negative imbalance, for example, in the early 1980s. Willis et al. (2004) used satellite altimetric height combined with about 900,000 in situ ocean temperature profiles to produce global estimates of upper-ocean (upper 750 m) heat content on interannual timescales from mid-1993 to 2002 (see Figure 4-3). Willis et al. calculated a 0.86 ± 0.12 W m^{-2} warming rate averaged over this period, but with large interannual variability. As seen in Figure 4-3, the ocean warming occurred in the later years of the record with little change in globally averaged ocean heat content prior to 1997.

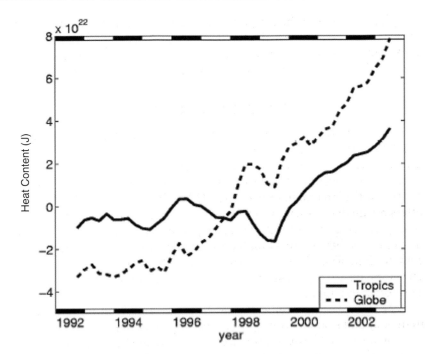

FIGURE 4-3 Interannual variability in upper-ocean (upper 750 m) heat content integrated over the region from 20°N to 20°S (solid line) and over the entire globe (dashed line). SOURCE: Willis et al. (2004).

5

Uncertainties Associated with Future Climate Forcings

Climate forcings are bound to evolve over the coming decades due to anthropogenic emissions as well as land-use changes. Scientists have developed methods for projecting future emissions and land-use changes, but limitations in these approaches lead to uncertainties in projections of future climate. Variability in natural processes—including the Sun and the Earth's orbit around it, volcanism, and internal variability in the climate system—can be even more difficult to project into the future. Nonetheless, making projections of how all of these forcings may manifest themselves enables policy analysts and decision makers to prepare accordingly. In this chapter, current capabilities for projecting future forcings are discussed and critical uncertainties associated with these forcings and their effect on climate are identified.

To date, all model projections of future climate have included a subset of climate forcings, typically greenhouse gas emission scenarios, solar variability, and more recently, aerosol emissions. As the diverse types of radiative and nonradiative climate forcings are recognized (e.g., aerosol indirect effects, changes in land cover), skillful projections of future global and regional climate will need to take them into account, an increasingly challenging task (Pielke Jr., 2001). Addressing this challenge may require a greater focus on assessing key societal and environmental vulnerabilities (Sarewitz et al., 2000).

FUTURE EMISSIONS OF GREENHOUSE GASES AND AEROSOLS

Estimates of future emissions of greenhouse gases, aerosols, and their precursors are fundamental for all aspects of climate change analysis. Emissions data are needed to drive climate models of the atmosphere, cryosphere, oceans, and biosphere. Scenarios of human-caused emissions are central to policy analysis of climate change because they provide future baseline trends for emissions that policy seeks to alter. The major driving forces of future emissions, such as demographic patterns, economic development, and environmental conditions, also underpin the assessment of vulnerability and development of adaptation strategies.

The available approaches for estimating future emissions to the atmosphere from human activities are described in this section. The methods used to relate emissions to atmospheric concentrations of greenhouse gases and aerosols are discussed in Chapter 6. A category of emissions that is not addressed in detail is volcanic eruptions. Estimating future volcanic emissions is not possible due to limitations in the understanding of the causes of volcanic eruptions, whose frequency depends on the rates of plate motions and of formation of large igneous provinces (see Chapter 3). One can only say, based on past history, that large climate-impacting eruptions will most likely occur, but it is not possible to predict when.

Anthropogenic Emissions

Projecting future emissions differs from other types of prediction that scientists make. Many natural systems, such as planetary motions, are governed by well-understood physical natural laws. The ability to predict future behavior and even events can be impressively high. Other physical systems may have well-understood physical laws, but these laws have shown that long-term prediction is impossible (e.g., chaotic systems). A third category includes systems to which the concept of governing laws expressed by mathematical equations is not applicable (Gaffin, 2002). For these systems, the driving forces change over time, sometimes radically, making prediction of specific outcomes a speculative effort. Many social, political, and economic science systems would fall into this category. Future anthropogenic emissions are a function of such socioeconomic systems and, as a result, are inherently unpredictable. Instead, emissions analysts offer "scenarios" that illustrate possible pathways for future emissions. These scenarios are not ruled out by current understanding of the driving forces behind emissions (Nakićenović et al., 2000). Figures 5-1 and 5-2 show scenarios for carbon dioxide (CO_2) and sulfur dioxide (SO_2) emissions.

In developing scenarios, it has been standard practice to consider futures with a range of policy interventions. On one end of the spectrum are noninterventionist emissions scenarios, which have been variously named business as usual, normative, reference, and no climate policy, among other terms. These scenarios seek to answer the policy question, How would

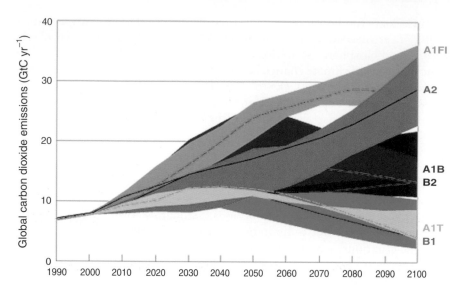

FIGURE 5-1 Global CO_2 emissions (billion tons of carbon per year) from all sources for the four scenario families (A1, A2, B1, B2) from the Intergovernmental Panel on Climate Change (IPCC) Special Report on Emissions Scenarios (SRES). The four scenario families had different assumptions about population, technological and land-use changes, and gross domestic product (GDP) growth rates. Each colored band represents the range of results for each scenario from the six emissions models used in the SRES report. Within any one scenario family, the emissions models all used the same scenario-specific global input assumptions for population and GDP growth. SOURCE: Nakićenović et al. (2000).

climate be affected if society takes no further policy steps to stem emissions and allows the driving forces behind emissions to take their course? It appears that none of the names are fully satisfactory and without ambiguities (Nakićenović et al., 2000). For example, "business as usual" does not convey that scenarios address economic and technology changes over a century time frame. Noninterventionist, the currently preferred term, also requires clarification because society is currently taking some steps, such as vehicular mileage standards and hybrid car marketing and research, that could be interpreted as climate policy.

Emissions scenarios that take account of policy intervention include those in which greenhouse gas concentrations are stabilized and cost-benefit scenarios. Stabilization scenarios are associated with atmospheric targets that policy makers often consider as long-term goals (IPCC, 1996;

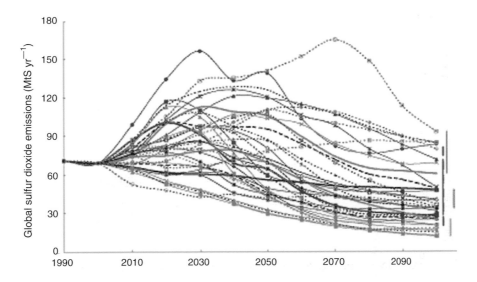

FIGURE 5-2 Global SO_2 emissions (million tons of sulfur per year) for the four SRES scenarios families and from each of the emissions models. Red lines correspond to the A1 family of scenarios, brown lines to A2, green lines to B1, and blue lines to B2. These emissions, along with the ozone precursor emissions of methane, carbon monoxide, nitrogen oxides, and volatile organic compounds, were also gridded onto $1° \times 1°$ latitude-longitude maps for use in global climate, atmospheric chemistry, and air quality models. The grids were developed using simple scaling applied to a base-year 1990 map and using the projected four to six regional changes from the SRES emissions models. SOURCE: Nakićenović et al. (2000).

Workshop on GHG Stabilization Scenarios, 2004). Such targets can easily be inverted by carbon cycle models to determine the required emissions scenario for stabilization (e.g., Wigley, 1991). Cost-benefit scenarios are another approach. They are based on the premise that the most efficient policy intervention is that in which the marginal cost of reducing emissions is balanced by the marginal reduction in climate damages, measured in monetary terms (Cline, 1992).

Developing emissions scenarios is tantamount to asking how different societies will produce, transform, and consume energy; extract and use Earth's resources; and modify the landscape for the next century. The possible answers to this vast and complex question are manifold. Uncertainties arise in all facets of the problem of building long-term scenarios. Moreover, unforeseen events, such as a revolutionary breakthrough in technology or a geopolitical shift, could occur and radically alter future emissions.

Methods for Developing Emissions Scenarios

Most emissions scenarios are developed using the IPAT model (Ehrlich and Holdren, 1971) in which environmental impact is the result of a multiplication of three driving forces: population, affluence per person, and technological impact per unit of affluence. When applied to greenhouse gas emissions, the impact is the rate of greenhouse gas emissions, while the technological factor is the rate of greenhouse gas emissions per unit of gross domestic product (GDP). The IPAT model has a long history in environmental studies, and there has been much debate over whether it is the proper approach (Chertow, 2001). As a purely mathematical multiplicative identity, it must yield correct emissions rates *if* all of the PAT factors are well known.

For example, world population growth today is approximately +1.3 percent per year (UN, 2002). Per capita GDP varies widely over time and by country, but recent estimates of world GDP per capita growth are approximately 1.1 to 1.4 percent per year (Maddison, 1995; World Bank, 2004). CO_2 emissions per unit of world GDP output display a well-documented long term decline, persistent over the past century, of about −1.3 percent per year (Nakićenović, 1996). The IPAT model indicates that adding these three rates should yield CO_2 emissions growth rates. The sum is about 1 to 1.5 percent per year growth in world CO_2 emissions, and this is what is seen, albeit with significant interannual variations, when using estimates of CO_2 emissions inventories (CDIAC, 2004).

Although the IPAT approach is reliable for estimating present-day emissions—as it must be if given good estimates for the main driving forces—using it to develop estimates of future emissions is by no means so easy. In principle, given a range of long-term projections for population, economic growth and technological changes, one could combine such projections to yield a range of future emissions rates. Some aspects of this approach are discernible in the predecessor scenarios to the Intergovernmental Panel on Climate Change (IPCC) Special Report on Emissions Scenarios (SRES) report: the IPCC IS92 emissions scenarios (Leggett et al., 1992).

Without doubt, the most sophisticated long-term projections exist for world and regional population changes, because these have been an area of intensive demographic research for many decades (Bongaarts and Bulatao, 2000; Lutz, 1996). Projections are routinely made for the world, regions, and countries by the United Nations (UN, 2002) and the International Institute for Applied Systems Analysis (IIASA; Lutz et al., 2004; Lutz, 1996). Less formal, or unpublished, projections are also developed by the

World Bank and the U.S. Census Bureau (Gaffin, 1998). The UN and the IIASA world and regional projections extend out to years 2150 and 2100, respectively. They include variants that explore the sensitivity of the projections to high and low fertility rate and mortality rate changes (and, to a more limited extent, changes in migration rate) that occur with development.

Although projections of future population are thus readily available for use in emissions models, long-term projections for the other key socioeconomic factors that drive emissions are generally not available. Economic and technological projections are much more tenuous and thus are usually limited to a few decades. As two examples, long-term projections for U.S. economic growth by the Energy Information Administration, possibly a best-country case, are made to 2025 (Annual Energy Outlook, 2004). The International Energy Agency makes world energy and economic projections out to 2030 (World Energy Outlook, 2004). Such time frames fall far short of those needed for climate and emissions scenarios.

To circumvent the relative dearth of long-term economic projections, the IPCC SRES report adopted a scenario approach in which story lines were developed that sketched in some detail the broad outline of future world development pathways. These story lines facilitated the selection of specific targets for economic growth in the year 2100. Four broad families of scenarios with story lines were developed and code-named A1, A2, B1, and B2. The A1 and A2 scenarios were characterized by an emphasis on economic and market forces within a geopolitical landscape of either strong globalization (A1) or a tendency toward regional development (A2). The B1 and B2 scenarios instead explored futures that were characterized by a pronounced value-shift toward environmental protection and sustainability, with a similar geopolitical dichotomy between globalization (B1) and regionalization (B2).

Within this framework, target GDP levels (in 1990 U.S. dollars) for the year 2100 were selected as follows: $550 trillion for the A1 family, $250 trillion for the A2 family, $350 trillion for the B1 family and $250 trillion for the B2 family (Nakićenović et al., 2000). Similarly, the exogenously available long-term population projections from the United Nations and the IIASA were also associated with each family based on the story line characteristics. The IIASA low fertility rate and low mortality rate projection was associated with the A1 and B1 scenarios, and the IIASA high-fertility and high mortality rate projection was associated with the A2 scenario (Gaffin, 1998). The UN medium projection was associated with the B2 scenario family. Each of the six emissions models used these exogenous inputs for the economic and demographic drivers.

Projecting technological evolution, including energy technologies that dominate emissions, is a similarly challenging undertaking involving consideration of many factors (Grubler et al., 1999; Nakićenović et al., 2000; Schipper, 1996). Integral to such projections is the observation that the world's economies and energy systems have been autonomously "decarbonizing" for centuries (Nakićenović, 1996; Nakićenović et al., 1998), that is, reducing the amount of carbon emitted per unit of economic output, primary energy produced, or both. This is happening for many reasons including societal preferences for improved air quality; changes in regional energy resources; and improvements in technology, energy efficiency, and productivity. Moreover, structural changes, as the world's economies move from preindustrial through industrial and into postindustrial phases, are leading to a general transition away from agriculture and manufacturing activities towards more service- and information-based sectors of activity (Soubottina, 2004). All of these changes contribute to decarbonization.

Long-term technology scenarios applied to emissions implicitly or explicitly project this decarbonization trend forward in time with varying rates (Nakićenović et al., 1998). A further uncertainty is that revolutionary advances in energy technologies or, alternatively, societal disasters are always possible, especially over the long term; however, such scenarios were eschewed in the most recent IPCC SRES report.

Even when long-term projections are available, the context in which they were developed may not be appropriate for noninterventionist emissions scenarios. Moreover, the emissions driving forces should not simply be combined in a random manner since fundamental interrelationships exist between them that may dictate the plausible nature of some combinations. For example, demographic, human welfare, and economic development are often seen as linked through various factors.

When the IPAT model is applied to greenhouse gas emissions, analysts have found it useful to further disaggregate the technology factor into the product of "emissions per unit of energy" (CO_2/E) and "energy per unit of GDP" (E/GDP). This variant of the IPAT formula is referred to as the "Kaya" identity (Kaya, 1990). The value added of disaggregating this way is that the CO_2/E factor is conveniently thought of as the carbon or greenhouse gas intensity of the energy system within a country, closely reflecting the fuel systems in use. The E/GDP can be thought of as the energy intensity of that economy. Because of persistent decarbonization, CO_2/E and E/GDP have declined for most economies and have been declining for perhaps two centuries (Nakićenović, 1996). For detailed modeling, structural models for the explicit energy, technological, economic, and demographic forces involved, developed for disaggregated regions, are preferable. Two broad categories of structural models are macro-economic, or "top-down," models (Nordhaus, 1993) and systems engineering, or "bottom-up," models (Nakićenović et al., 2000). Top-down models focus on developing large-scale economic production functions involving capital stock, labor productivity, and broad-scale technological change. The production function requires decision variables for investment and savings rates. The model seeks an optimal economic growth rate that maximizes utility. The strength of these models is their treatment of the interactions of the economy as a whole. Their weakness is the lack of explicit simulations of energy technologies. In contrast, bottom-up models focus more explicitly on the energy technologies in use. They include details about primary and secondary energy production, distribution, and end-use consumption, seeking to balance energy supply and demand by adjusting energy prices. There are models that combine the top-down and bottom-up approaches (e.g., Goldstein and Greening, 1999). In anticipation of its Third Assessment Report (TAR), the IPCC developed a new set of emissions scenarios. This effort resulted in the *IPCC Special Report on Emissions Scenarios* (Nakićenović et al., 2000), often referred to as SRES, which is the most recent comprehensive effort at developing projections of future anthropogenic emissions. The results have been widely used in the IPCC TAR and other climate modeling applications. Both top-down and bottom-up modeling approaches are represented in the SRES (Nakićenović et al., 2000). Details on the individual model components are discussed in the SRES and the source references (deVries et al., 1994; Morita et al., 1994; Messner and Strubegger, 1995; Edmonds et al., 1996a,b; Pepper et al., 1998; Mori and Takahashi, 1999).

Since the release of the SRES, emissions scenarios have continued to evolve. For example, the declines in fertility rates since the SRES have been incorporated in more recent work, as has the potential for longer human lifetimes. Another important development is the incorporation of explicit energy and agriculture technologies in models, bridging the gap between economics and engineering. Integrated assessment models now routinely produce scenarios with carbon capture and storage. Some models are now explicitly modeling advanced transportation technologies such as hydrogen, biofuels, hybrids, diesels, and advanced internal combustion engines. Similarly, some models are now producing scenarios that consider the competition between land uses implied by large-scale development of commercial biomass crops and the interaction between the energy system, land use, and the terrestrial carbon cycle.

Emissions Scenarios for Short-Lived Compounds

One category of future emissions with particularly high uncertainties is the shorter-lived compounds, including tropospheric ozone precursors (carbon monoxide [CO], nitrogen oxides [NO_x], methane [CH_4], and volatile organic compounds [VOCs]), sulfur dioxide, carbonaceous aerosols, dust, and sea salt. Ozone and aerosol concentrations are highly variable over space and time; thus, the spatial distribution of emissions is important for accurately simulating their atmospheric distributions and radiative impacts. The present-day source inventories for these gases are complex and need further work.

The IPCC SRES provided estimates for emissions of methane, other ozone precursors, and sulfur dioxide (precursor for sulfate aerosols), but not for carbonaceous aerosols, dust, or sea salt (Nakićenović et al., 2000). The SRES paid special attention to sulfur (Grubler, 1998), motivated by the burgeoning recognition between the IPCC's First and Second Assessment Reports of strong radiative forcing of sulfate aerosol (IPCC, 1996). Most of the SRES effort was devoted to estimating temporal trajectories for large geographic regions—a significant challenge for 100 years into the future—and less attention was paid to smaller-scale spatial variability. In the SRES models, future emissions of species with poorly inventoried sources were often estimated using simple relationships to aggregate energy (e.g., proportional to fossil fuel combustion), economic indicators, and sector-specific activities (e.g., agriculture). Providing more detailed spatial projections for short-lived emissions remains an important challenge for analysts.

To fill the currently burgeoning need for present and future estimates of black carbon and organic carbon emissions, the IPCC (2001) made nominal guesses at these emissions by applying poorly known scaling factors to the SRES CO emissions. Possible future changes in the relative contributions of biomass and fossil fuel combustion to CO emissions were not evaluated. IPCC (2001) noted that this could bias estimates of future particulate emissions if air quality policies target them differently from CO. Explicit, gridded carbonaceous aerosol emissions scenarios that consider a range of policy outcomes are needed.

Spatial Variability in Emissions

As a first step in providing spatially resolved emissions, SRES used a top-down approach. Emissions data were prepared for four large aggregated regions: (1) the Organization for Economic Co-operation and Development (OECD) 1990, (2) Asia, (3) Eastern Europe and the former Soviet Union, and (4) the rest of the world. The six emissions models used in the SRES each had a more detailed aggregation with 9 to 13 regions; each of these models subsequently reaggregated the individual regional data to match the four reporting regions. Results for the four regions are presented in the final report for all of the long-lived greenhouse gases. For the short-lived gases, SRES scaled the aggregated projections from the four reporting regions (SO_2 used six regions) to a gridded map for 1990. The base-year emissions map was from the Emissions Database for Global Atmospheric Research (EDGAR), maintained at the National Institute of Public Health and the Environment in the Netherlands (Olivier et al., 1996; Van Aardenne et al., 2001; see *http://arch.rivm.nl/env/int/coredata/edgar/intro.html*). This procedure is a reasonable first cut at a complex problem, although it does not capture temporal trends in the spatial distribution of emissions.

One area for improvement is to include future population change information at as high a spatial resolution as possible. The EDGAR database used 1990 population distribution as a major part of its spatial partitioning of emissions to a 1° latitude by 1° longitude map (Van Aardenne et al., 2001). For example, 1990 population density was used to spatially allocate all fossil fuel emissions from power plants; residential and industrial energy use; transportation; industrial and domestic biofuel combustion; processing iron, steel, and cement; and solvent production. By rescaling the base-year grid, therefore, SRES essentially perpetuates the 1990 world population distribution throughout the forecast period to 2100. The SRES used 100-year regional population projections from the UN and the IIASA as one of the key drivers behind emissions (Gaffin, 1998). However, this does not rectify the spatially static character inherent in rescaling the 1990 EDGAR emissions grid because only the reporting region emissions results were used in the top-down gridding, rather than explicit future population distribution information.

To capture future spatial variability in source emissions better, the research community should consider more explicit methods, such as bottom-up methods, for including key indicators. For example, population projections exist nationally out to 2050 from the UN and other sources (UN, 2002). These national projections should be incorporated in some fashion into future gridding algorithms for emissions locations. In addition, there are numerous subnational population projections for major emitting nations, including the United States (U.S. Census Bureau, 2002) and, recently, China (Toth et al., 2003) and India (Srinivisan and Shastri, 2004). Subnational projections, although difficult to undertake, would be valuable for emissions scenarios. Newer sources of remotely sensed data on land use and anthropogenic activity also provide valuable spatial data. For example, nocturnal lighting data are of high quality and can be closely correlated to a number of anthropogenic indicators including emissions (Elvidge et al., 1997). Given that many of the emissions sources are poorly quantified, nocturnal lighting data should be mined for possible improvements in the inventories. Such data offer alternative spatial baselines upon which projected grids could be based (Plutzar et al., 2000).

Biogenic VOC Emissions

Volatile organic compounds are important ozone precursors. In addition to anthropogenic sources, they are emitted by vegetation at a rate that depends on temperature, radiation, and other factors. Biogenic VOC emissions are expected to be highly sensitive to climate change, but the dependences are complicated. The dependence on temperature is large, amounting to a doubling of emissions for a 7 K temperature increase (Guenther et al., 2000). The dependence on solar radiation is weaker and also relatively well understood. Increasing CO_2 is known to cause reductions in isoprene emissions (Sanadze, 1969; Monson and Fall, 1989), although this may be compensated by increases in the vegetation leaf area index associated with CO_2 fertilization. Most uncertain is the response of biogenic VOC emissions to changes in vegetation type, as driven both by anthropogenic land use and by climate change. Quantifying this effect requires vegetation dynamics models, but these are difficult to constrain for future projections.

Lightning NO_x Emissions

Thermal and ionic chemistry taking place in lightning strokes produces large amounts of NO_x from nitrogen (N_2) and oxygen (O_2). Lightning NO_x emissions are likely the most important precursors of ozone in the upper troposphere (Martin et al., 2002; Li et al., 2004) and also the most uncertain of the major NO_x sources (Nesbitt et al., 2000; Tie et al., 2002). Credible global estimates from the literature range from 2 to 20 Tg of nitrogen per year (IPCC, 2001). Process studies of lightning emissions based on flash energies tend to be at the high end of that range (Price et al., 1997), while estimates constrained by atmospheric observations of nitrogen oxides and ozone tend to be at the low end (Wang et al., 1998; Martin et al., 2002; Staudt et al., 2003). Lightning emissions could possibly be greatly enhanced in populated regions due to high aerosol emissions (Steiger and Orville, 2003).

Dust and Biomass Burning Emissions

Several studies have applied land-use models to estimate past, current, and future dust emissions (Matthews, 1983; Alcamo et al., 1994; IPCC, 2001; Mahowald and Luo, 2003). Mahowald and Luo (2003) found an 11 percent increase in dust emissions over the last century (from 1900 to 2000) but predicted a 20 percent decrease in the next century (from 2000 to 2100). Ice core measurements of dust concentrations over the last century are inconclusive. The projections of future decreases in dust emissions result from changes in soil moisture and surface winds in arid regions; however, future human landscape conversion and landscape degradation have not been included in these projections. Decreases could reduce cooling, thereby enhancing climate warming trends. More information is needed to constrain predictions of dust emissions for the twenty-first century and beyond.

The future of biomass burning emissions is also not well understood. Current evidence shows increasing rates of biomass burning (Cochrane et al., 1999). Additional work is in progress to better understand the current and future sensitivities of biomass burning to biological and societal driving forces (Lavorel et al., 2005; Spessa et al., 2003).

Climatic Implications of Air Quality Regulations

Emissions of aerosol and ozone precursors are heavily regulated in developed countries because of their impacts on air quality. These regulations will likely become stronger in the future as air quality standards are restricted. Control strategies have not yet been developed in a manner that addresses both air quality and the potential climatic implications of emissions reductions (NRC, 2005).

Controlling emissions that lead to air pollution can have either warming or cooling effects on climate, as shown in Table 5-1. Regulations targeting black carbon emissions or ozone precursors would have combined benefits for public health and climate (Hansen et al., 2000; Jacobson, 2002).

TABLE 5-1 Responses of Regional Haze and Climate to Reductions in Emissions of Aerosols and Aerosol Precursors

Reduction in Pollutant Emissions	Impact on Regional Haze[a]	Impact on Aerosol Direct Effect[b]
SO_2	↓	↑
NO_x	↑↓ [c]	↑↓
VOC	↑↓ [c]	↑↓
Ammonia (NH_3)	↓	↑ [d]
Black carbon	↓	↓
Primary organic compounds	↓	↑
Other primary particulate matter (crustal, metals, etc.)	↓	↑

[a]Direction of arrow indicates increase (↑) or decrease (↓), and color signifies undesirable (red) or desirable (blue) impact; size of arrow signifies magnitude of change. Small arrows signify possible or small change.

[b]Indirect effects through clouds and precipitation are highly uncertain. Note that the extent and possibly the scale of climate impacts for listed pollutants are quite different from CO_2 and CH_4. Direction of arrow indicates warming (↑) or cooling (↓).

[c]No change if little NH_3 is available in atmosphere.

[d]More accurately, decreased aerosol-induced cooling.

SOURCE: NARSTO (2003).

BOX 5-1
Implications of Aerosol Forcing for Climate Sensitivity

Greenhouse gas forcing increased by about 2.1 W m^{-2} since 1860 (Feichter et al., 2004), and the reported surface warming trend was about 0.6 K (IPCC, 2001). Some models that have been used to improve understanding of the relationship between the forcings and climate response have included only greenhouse gases, which lead to a warming. Others have included greenhouse gases, which lead to a warming, and aerosols, which lead to a cooling. As a thought experiment, consider two models that accurately simulate the observed temperature trend. Model A, which does include aerosols, estimates that it took the entire 2.1 W m^{-2} greenhouse gas forcing to produce the 0.6 K warming. On the other hand, Model B includes aerosols that offset 50 percent of the greenhouse gas warming with cooling. The net forcing increase in Model B is thus only 1.05 W m^{-2} (50 percent of 2.1 W m^{-2} contributed by the greenhouse warming). In Model B, it therefore takes only 1.05 W m^{-2} to produce the observed 0.6 K warming. If aerosol cooling were suddenly removed from Model B (e.g., aerosol emissions were significantly decreased in year 2005 due to their health impacts), it would lead to an additional warming during the next few decades without any further addition of greenhouse gases.

Methane reductions have been proposed as an effective double dividend for improving ozone air quality and mitigating climate change (Fiore et al., 2002). In the case of sulfate and other aerosols having a negative radiative forcing, reducing aerosol concentrations will likely create a more positive radiative imbalance and could actually accelerate increases in surface temperature. Because understanding of the effect of aerosols on the hydrological cycle and vegetation is still incomplete, it is difficult to predict the total effect on climate of reducing aerosol emissions.

Stricter regulation of scattering aerosols such as sulfate could also reveal potential problems with assumptions in climate models about the magnitude of climate sensitivity. Several modeling studies have attributed the surface temperature increases observed during recent decades to the addition of CO_2 and other greenhouse gases (IPCC, 2001). More recent modeling studies that have incorporated aerosols show that aerosol direct and indirect forcing may have offset as much as 50 to 75 percent of the radiative forcing due to greenhouse gas increases since preindustrial times. Models that do not include cooling by aerosols will tend to underestimate climate sensitivity, which is adjusted to produce the best fit with the observed radiative forcing and temperature trends (see Box 5-1). This evidence implies a greater possibility that the climate sensitivity for the radiative effects of a doubling of CO_2 is in the upper range of the 1.5 to 4.5 K global-averaged surface warming (Andronova and Schlesinger, 2001; Wigley and

Raper, 2001; Forest et al., 2002; Gregory et al., 2002; Anderson et al., 2003a).

SOLAR VARIABILITY

At present, the observational database is too short to detect long-term solar irradiance trends, nor is it possible to predict future solar activity with any skill. Empirical evidence suggests that the approximately 11-year activity cycle of the Sun might be expected to produce future changes in total solar irradiance of order 1 W m^{-2} or less. This is the change observed in the past three solar cycles, two of which are the second and third largest since 1610 (Lean, 2001).

Unknown is whether solar activity will increase or decrease in the future and how long-term secular changes, if they exist, might evolve. That current levels of solar activity are at overall high levels, according to both sunspot numbers and cosmogenic isotopes, suggests that future solar irradiance values may not be significantly higher than seen in the contemporary database. A projection of future solar activity based on spectral synthesis of the cosmogenic isotope record confirms that solar activity is presently peaking and in 2100 will reach levels comparable to those in 1990 (Clilverd et al., 2003). However, projections of combined 11-, 88-, and 208-year solar cycles suggest that an overall increase in solar activity from 2000 to 2030 will produce a top-of-the-atmosphere (TOA) climate forcing of +0.45 W m^{-2}, followed by decreasing activity until 2090 with climate forcing −1 W m^{-2} (Jirikowic and Damon, 1994). In contrast, a numerical model of solar irradiance variability that combines cycles related to the fundamental 11-year cycle by powers of 2 predicts a 0.05 percent decrease in irradiance during the next two decades (Perry and Hsu, 2000), for which associated climate forcing is −0.1 W m^{-2}. A lack of physical understanding of how the dynamo-driven solar activity produces the competing effects of sunspot blocking and facular brightening cautions against future predictions, even of 11-year cycle amplitudes.

LAND-USE AND LAND-COVER CHANGE

With continued population growth in the twenty-first century, the conversion of land into agricultural, urban, and other human uses will continue and could even accelerate. The dynamics of anthropogenic land-use change is quite complex and involves socioeconomic forcings and feedbacks (Lambin et al., 2003; Napton et al., 2003; Sohl et al., 2004), as well as climate forcings and feedbacks. For example, Figure 5-3 demonstrates how highway construction and other major infrastructure projects might contribute to change in Brazilian vegetation. The availability of satellite moni-

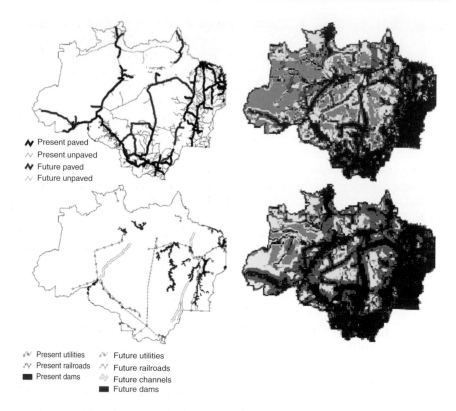

FIGURE 5-3 Left: Existing and planned highways and infrastructure projects in the Brazilian Amazon: *(top)* highways and road and *(bottom)* major infrastructure projects. "Utilities" are gas lines and power lines, while "channels" are river channelization projects. Right: Predicted forest degradation by the year 2020 with *(top)* "optimistic" and *(bottom)* "pessimistic" scenarios. Black is deforested or heavily degraded, including savannahs and other nonforested areas. Red is moderately degraded, yellow is lightly degraded, and green is pristine. SOURCE: Laurance et al. (2001).

toring of land-use change (e.g., Steyaert et al., 1997; Loveland et al., 2000, 2002) for the United States and globally will permit detailed data to be obtained into the future.

The consequences of future land-use change on the forcing of the climate system include alterations in the surface albedo, portioning of the surface net radiation between latent and sensible turbulent fluxes, and trace gas and aerosol fluxes. Although understanding of how land-use change acts as a climate forcing is not as developed as that for greenhouse gas forcing, a recent review of research on this topic concluded that land-use change during the twentieth century had well-defined effects on regional climate (Kabat et al., 2004). In some cases, these regional effects are estimated to be as large as those due to a doubling of CO_2. In addition, land-use change may have possible significant effects on the global climate through teleconnections. Continued conversion of the landscape is expected to have a comparable effect in the next century.

The climate forcing of future landscape change on the surface energy and water budget, and how this effect is transmitted through the atmosphere, has been little studied. Most previous work on future landscapes has concentrated on how the landscape responds to future climate change simulated in models (e.g., NAST, 2001) and on surface greenhouse gas sources or sinks, rather than on how the landscape itself alters climate through its heat and moisture fluxes (e.g., Snyder et al., 2004).

ABRUPT CLIMATE CHANGE

Abrupt climate changes take place when "the climate system is forced to cross some threshold, triggering a transition to a new state at a rate determined by the climate system itself and faster than the cause" (NRC, 2002). Paleoclimate records indicate that abrupt changes have occurred frequently in the past and at rates high enough that human or natural systems may have had difficulty adapting. However, efforts to develop model-based scenarios for future abrupt climate changes are limited by our lack of understanding of the mechanisms that underlie such changes (see NRC, 2002, for a full discussion of this issue).

One example is a striking series of abrupt changes in temperature identified in Greenland ice core records from the last glaciation. Known as Dansgaard/Oeschger (D/O) cycles, these temperature shifts recur on millennial time scales. They have been attributed to threshold jumps in thermohaline circulation (THC), presumably triggered by sudden discharges of icebergs and freshwater from ice sheets (Broecker, 1997). Temperatures above the ice cap may have dropped by as much as 8°C, and large parts of western Europe may have cooled by a few degrees. The recent discovery of correlatives of D/O cycles in tropical latitudes, however, has raised an alternative possibility that the underlying mechanism was ocean-atmosphere interaction in low latitudes (Clement et al., 2000). Some have even argued that El Niño cycles of long duration or shifts in the ITCZ (intertropical convergence zone) might change water vapor export to high latitudes. Such changes could have amplified or, through impacts on THC, even caused the D/O variations (Peterson et al., 2000).

Examples of abrupt change have been identified in the Holocene as well. Between about 12 and 5 thousand years ago the last phases of deglaciation led to repeated reroutings of glacial lake meltwater into the convecting regions of the North Atlantic (e.g., Teller and Leverington, 2004). It has been argued that each of these events weakened the North Atlantic THC enough to cause abrupt cooling of the northern North Atlantic, Greenland, Iceland, and Europe, perhaps by as much as a few degrees. Yet proxy data from the northern North Atlantic so far have revealed no compelling evidence for changes in THC (Keigwin and Boyle, 2000).

Even if, as many assume, threshold behavior in THC is the most likely cause of abrupt climate change, the processes that lead to such behavior cannot be quantified. The ways in which the North Atlantic THC interacts with other oceans is not well known (Whitworth et al., 1999); it is not certain whether freshwater forcing actually weakens THC under all conditions (Marotzke, 2000). Further, the way in which convective activity is linked to THC is not fully understood (Marotzke and Klinger, 2000).

An abrupt climate change would have more severe impacts on natural and human systems than more gradual change. In particular, it could result in more extremes in climate variables (e.g., heat index, storm intensities and frequencies). Knowledge of what triggers abrupt climate changes is still quite limited; more research is needed to determine whether human-caused increases in greenhouse gases or land-use changes might lead to an abrupt change. Indeed, past abrupt climate changes have been especially common when the climate system itself was being altered (NRC, 2002).

6

Research Approaches to Furthering Understanding

Previous chapters have covered the current understanding of radiative forcings, how the forcings have varied over Earth's history, different ways to quantify forcings, and critical uncertainties involved in predicting future forcings. This review of current understanding has illustrated that significant knowledge of forcings—including knowledge of their sources, magnitudes, variations, and effects on climate—has been achieved over the past decades and that there are still many critical unknowns. In this chapter, the many research approaches for studying forcings are described. These include observations from multiple platforms (e.g., surface observing networks, satellite-based remote sensing instruments), laboratory and process studies, atmospheric reanalysis and data assimilation, tools to relate emissions to atmospheric concentrations, "proxy" observations of past forcings and response, and a variety of climate modeling approaches.

OBSERVATIONS OF RADIATIVE FORCING AND RESPONSE

Robust observations of radiative forcings are critical for improving understanding of these climate drivers, how they varied in the past, and how they might change in the future. Current observational approaches include in situ and surface-based monitoring of greenhouse gases and aerosols; satellite-based observations of atmospheric composition, land cover, and solar variability; and intensive campaigns that utilize aircraft-based observations with in situ and satellite measurements to study processes in detail. Observations of climate response, such as surface temperature or ocean heat content, also provide important information about climate forcings. Much of the current understanding of radiative forcing and other forcing concepts has been obtained from climate models. To improve this understanding, routine observations of climate forcings will be essential, both as a record of change in the climate system and as a critical constraint for climate models.

Long-Lived Greenhouse Gases

The major long-lived greenhouse gases (carbon dioxide [CO_2], methane [CH_4], nitrous oxide [N_2O], and halocarbons) are all extensively observed by surface networks such as the National Oceanic and Atmospheric Administration (NOAA) Climate Monitoring and Diagnostics Laboratory (CMDL) and the Atmospheric Lifetime Experiment (ALE)/Global Atmospheric Gases Experiment (GAGE). All have sufficiently long lifetimes to be well mixed in the atmosphere. Their spectroscopy is also well established. Radiative forcings can thus be assessed with confidence.

There is, however, a strong impetus to improve the observational system for these gases in order to constrain inverse model analyses of their regional budgets. For example, many analyses have used the large-scale gradients of CO_2 measured from the surface networks to constrain the global carbon budget and quantify the terrestrial sink at northern midlatitudes. However, they have not been successful in determining the longitudinal distribution of the carbon sink among the three northern midlatitude continents. The International Geosphere-Biosphere Programme (IGBP) TransCom activity (*http://transcom.colostate.edu/*) has provided a forum for standardizing and comparing these inverse model analyses, but model transport errors ultimately limit their ability to exploit the relatively sparse surface air observations in terms of regionally resolved source and sink constraints (Gurney et al., 2002).

Better understanding of terrestrial uptake is critically needed for future projections of CO_2 concentrations (IPCC, 2001). An extensive network of CO_2 flux measurement towers has been deployed worldwide in recent years and is coordinated through the FLUXNET activity (Baldocchi et al., 2001). It includes in particular the AmeriFlux network in North America (*http://public.ornl.gov/ameriflux/*). These measurements provide direct observations of the terrestrial component of the carbon budget and also the biogeochemical constraints needed to interpret these observations. However, it has not been clear how to integrate them into large-scale inverse model analyses. The North American Carbon Program (NACP) outlines a strategy for doing so, involving in particular the use of aircraft observations to scale up the tower flux observations and providing a linkage to the global observation network (Wofsy and Harriss, 2002; Denning et al., 2003).

Global mapping of CO_2 concentrations from space would greatly improve our ability to constrain carbon sources and sinks in inverse models. It would pave the way for construction of national carbon budgets, providing important input for global environmental agreements aimed at mitigating climate change. The challenge is to deliver a measurement with sufficiently high precision to be useful for inverse modeling. A precision of 0.3 ppmv (parts per million by volume) is thought to be necessary (Pak and Prather, 2001; Rayner and O'Brien, 2001). The Orbiting Carbon Observatory (OCO) satellite instrument, planned for launch in 2007, is expected to provide this precision (Crisp et al., 2004). It will measure CO_2 column mixing ratios with kilometer-scale spatial resolution by solar backscatter in the 1.58 μm band, with measurements in additional bands to correct for aerosol and surface pressure effects. Simulations with chemical transport models sampled along the OCO orbit track suggest that the measurements should be of great value for constraining carbon fluxes down to a regional scale (Crisp et al., 2004).

Methane concentrations have increased by a factor of 2.5 since the eighteenth century, but the rate of growth began to slow in the 1980s and was close to zero in 1999-2002 (Dlugokencky et al., 2003). The reason for this slowdown is not clear. Changes in agricultural practices, decreased natural gas production in Russia, and increasing OH concentrations (reducing the lifetime of methane) may all have contributed (Khalil and Shearer, 1993; Dentener et al., 2003; Wang et al., 2004). A number of inverse model studies have been conducted to constrain sources of methane using long-term observations from the NOAA CMDL network (Hein et al., 1997; Houweling et al., 1999; Wang et al., 2004), but they do not yield consistent results. Aircraft observations in continental outflow over the northwest Pacific have been used recently to constrain Eurasian sources of methane (Xiao et al., 2004) and halocarbons (Palmer et al., 2003). Satellite measurements of methane and halocarbons have so far been restricted to the stratosphere. There has been great interest in using solar backscatter measurements to constrain the column mixing ratio of methane (Edwards et al., 1999), but efforts so far have been unsuccessful. Similar to CO_2, satellite observations of methane with sufficiently high resolution would increase considerably our ability to constrain regional sources.

Ozone

Ozone has a lifetime ranging from days to months in the troposphere and up to years in the lower stratosphere. Its distribution in the atmosphere is thus highly variable, in contrast to the long-lived greenhouse gases. Vertical profiles from ozonesondes provide at present the best characterization of the global distribution of ozone. Their coverage is extensive in the extratropical Northern Hemisphere but relatively sparse in the tropics and the

Southern Hemisphere. Relatively low sampling frequencies (typically weekly) and calibration issues have made it difficult to use these observations to quantify long-term trends of ozone and its vertical distribution, in both the troposphere and the stratosphere (Logan, 1999). This uncertainty in ozone trends and our ability to describe them in models is the main difficulty in quantifying the radiative forcing of ozone in the past and making projections for the future.

A global climatology of total ozone columns extending back to 1979 is available from the Total Ozone Mapping Spectrometer (TOMS; see Figure 6-1) and other sensors, and has been used extensively and successfully for trend analyses (WMO, 2003). A similarly long, although sparser, record is available for the vertical distribution of ozone down to the lower stratosphere from the Stratospheric Aerosol and Gas Experiment (SAGE) and

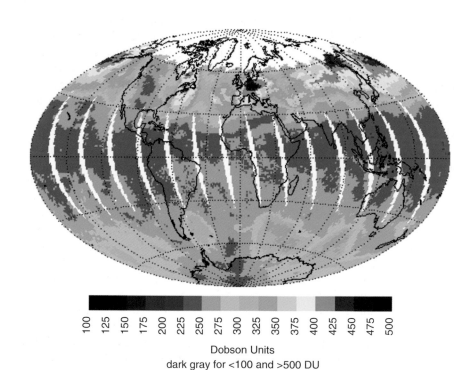

Dobson Units
dark gray for <100 and >500 DU

FIGURE 6-1 Total column ozone observed on January 25, 2005, by the Total Ozone Mapping Spectrometer (TOMS) aboard the Earth Probe satellite. SOURCE: NASA Goddard Space Flight Center.

other sensors. Most problematic are the tropopause region and the troposphere, which are of most interest from a radiative forcing standpoint. Despite these limitations, for this short-lived forcing, unlike for other such species, chemical transport models are not needed to evaluate the forcing because of the presence of a reliable, continuous global monitoring network.

The inadequacy of current tropospheric ozone observations for constraining global distributions and trends has spurred the concept of an Integrated Global Atmospheric Chemistry Observation System (IGACO) to integrate and expand the current observational network (Barrie et al., 2004). Satellite observations have to play a key role in mapping the global distribution. There is at present no direct measurement of tropospheric ozone from space. A number of attempts have been made to constrain tropospheric ozone columns from a combination of independent measurements of the total column and the stratospheric contribution, starting from the pioneering work of Fishman et al. (1990), but there are large uncertainties with these products even at equatorial latitudes where they are most robust (Martin et al., 2002). Some attempts have been made to infer tropospheric ozone columns from solar backscatter measurements, but the results so far are only qualitative. The Tropospheric Emission Spectrometer (TES), launched on the Aura satellite in July 2004, will provide the first opportunity for global mapping of tropospheric ozone from space. It will observe infrared emission of ozone in the nadir and in the limb with line-by-line resolution (Beer et al., 2001). Algorithm development studies suggest that it should provide one to two constraints on the vertical profile in the troposphere with sufficient precision to allow global mapping (Clough et al., 1995; Bowman et al., 2002).

Aerosols

Observational approaches to better understand aerosol radiative forcing include closure studies, remote sensing from space-based and other platforms, Lagrangian studies, and surface-based observations, which are described in more detail below. Until recently, models were needed to infer the direct forcing. However, recent field campaigns, including the Indian Ocean Experiment (INDOEX) and the Aerosol Characterization Experiment in Asia (ACE-Asia), have obtained the direct forcing from radiation budget observations at the surface and the top of the atmosphere (TOA; Figure 6-2). Regarding direct radiative forcing by tropospheric aerosols, there are several tests that have been and should continue to be performed between models and existing observations. These include comprehensive comparisons against surface concentration measurements, aerosol optical depth measurements (e.g., AERONET), reflected radiation flux at TOA

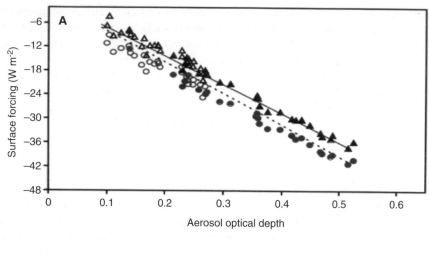

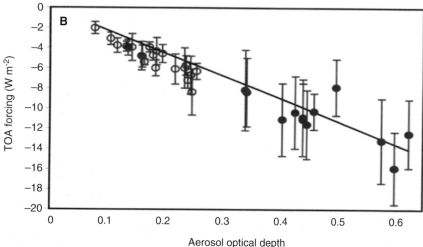

FIGURE 6-2 Direct observations of clear-sky forcing efficiency. The top panel shows the reduction at the surface due to aerosols as a function of the aerosol optical depth, while the bottom panel shows the same at the top of the atmosphere. Surface measurements were obtained with broadband pyranometers and spectral radiometers, while the TOA measurements were obtained from Clouds and the Earth's Radiant Energy System (CERES) radiation budget instruments on board the Tropical Rainfall Measuring Mission (TRMM) satellite. The values are diurnal mean values for clear skies and the data are for the Arabian Sea. The figure shows that the surface forcing is three times the TOA values, both being negative. The forcing includes both natural and anthropogenic aerosols, and the measured single scattering albedo is about 0.9 (±0.03). SOURCE: Satheesh and Ramanathan (2000).

(e.g., Earth Radiation Budget Experiment [ERBE] clear-sky measurements over oceans), radiation measurements at the surface (e.g., Baseline Surface Radiation Network [BSRN]), and vertical profiles where available. Long-term monitoring is essential to understand interannual variations in forcing by short-lived species. Finally, extracting the indirect effect from observations, particularly those based on regional and global datasets, may require one to deal with the response of cloud systems to the thermodynamic environments that are tied to the polluting particles (Harshvardhan et al., 2002).

Closure Studies

Closure experiments provide constraints on aerosol radiative properties. In a closure experiment, an aerosol property is measured in one or more ways and then calculated from a model based on independently measured data (Quinn and Coffman, 1998). The objective is to evaluate models using a collection of independent observed quantities to provide multiple constraints on the aerosol properties being analyzed. Closure studies of aerosol direct and indirect effects typically use multiple measurements in a single atmospheric column at one moment in time to constrain the radiative forcing. The comparison between the calculated and measured values provides a test for the reliability of the measurements and the model.

Successful closure experiments have been conducted in a number of field campaigns including ACE-1 and ACE-2, the Tropospheric Aerosol Radiative Forcing Observational Experiment (TARFOX), INDOEX, and ACE-Asia. For example, ACE-1 and ACE-2 provided detailed aerosol characterization that showed good agreement between modeled and observed optical depth (Quinn and Coffman, 1998; Collins et al., 2000; Redemann et al., 2000; Russell et al., 2000; Fridlind and Jacobson, 2003; Wang et al., 2003). Closure experiments were conducted using a collection of vertically resolved measurements of aerosol size and composition with simultaneous vertical profiles of spectrally resolved optical depth. Light scattering and one-dimensional radiative transfer calculations were then used to calculate the optical depth profile, and these calculated values were compared with the aerosol size and composition-based calculations of optical depth.

Other closure experiments have provided important constraints on the direct effect of aerosols on radiation. Collins et al. (2000) thus determined that multiple aerosol layers in the atmosphere are significant in scattering light in the troposphere, showing a good correspondence between measured aerosol concentrations and measured scattering. It appears from these and other aircraft-based closure experiments that aerosol forcing is well understood when the column loading and the distribution of aerosol size and composition have been characterized.

Integrated Approaches for Obtaining Aerosol Forcing from Observations

Since 1999 there have been several successful efforts to obtain aerosol radiative forcing information from surface, aircraft, and satellite observations (e.g., Figure 6.2). The success of these studies clearly illustrates the need for accurate observations of radiation budget, aerosol optical depth, and cloud fraction and cloud type at the surface (in selected regions) and from space. In situ aerosol chemical data from aircraft have been used to separate anthropogenic from natural forcing. Surface-based aerosol column optical measurements have been combined with Moderate Resolution Imaging Spectroradiometer (MODIS) data for clear skies to obtain aerosol forcing at the top of the atmosphere (Kaufman et al., 2002). These clear-sky forcing values provide an important constraint on the closure approaches described earlier and on climate model simulations of direct aerosol forcing. Integrated approaches have also been very effective in capturing the effect of aerosols in nucleating more cloud drops and in suppressing precipitation efficiency. For example, the spectral dependence of cloud reflectivity measured from space has been used to obtain the effective radius of clouds (e.g., Coakley et al., 1987; Nakajima et al., 2001). Comparisons of the effective radius between pristine and polluted clouds have provided estimates of the global indirect effect, although additional work is needed to improve the accuracy of these estimates. In situ aircraft observations have been used to characterize the dependence of cloud drop number density and effective radius on aerosol number concentration and cloud condensation nuclei (CCN) for low clouds (Taylor and McHaffie, 1994; Gultepe et al., 1996; Pawlowska and Brenguier, 2000; McFarquhar and Heymsfield, 2001) and high clouds (Sherwood, 2002). Satellite data for aerosol optical depth and cloud fraction have been used to infer the semidirect effect (Koren et al., 2004; Kaufman and Fraser, 1997). Major new insights into the role of anthropogenic aerosols in reducing precipitation efficiency have been obtained by combining satellite data for effective cloud drop size, precipitation rate (using microwave radiometer and radar), and aircraft data (Rosenfeld, 1999, 2000).

To exploit the new generation of satellite data for clouds and aerosols (e.g., the National Aeronautics and Space Administration [NASA] A-Train), in situ aerosol-cloud observatories are needed in different regions of the planet (preferably the regions contributing most to anthropogenic aerosol forcing). This combination of satellite and in situ data will enable us to address fundamental issues, including the global distribution of black carbon; regional statistics of aerosol number concentration, composition, CCN, and cloud drop distribution; and global distribution of aerosol forcing at the surface and the TOA.

Lagrangian Studies of the Indirect Effect

The critical issue in field studies addressing the indirect aerosol effect is that simultaneous measurements of the aerosol entering the cloud and of the cloud microphysical characteristics are needed. A Lagrangian sampling strategy is essential. This approach was tried during ACE-2 with limited success due to the complexity of regional boundary layer dynamics that resulted in particularly complex clouds and decoupled mixed layers (Johnson et al., 2000; Sollazzo et al., 2000). Ship tracks have provided an opportunity for Lagrangian sampling and yielded evidence that even hydrophobic organic compounds may be incorporated in cloud droplets (Russell et al., 2000).

Surface-Based Observations

Surface sites and ships provide platforms for long-term continuous measurements. Ground-based experiments have studied the role of cloud-particle interactions through fog events and showed that chemical composition is a key factor in determining cloud droplet activation properties (Noone et al., 1992). Recent studies have shown evidence consistent with activation of organic particles (Facchini et al., 2000; Decesari et al., 2001; Ming and Russell, 2004). Additional long-term datasets at surface sites may provide statistically significant constraints on direct and indirect aerosol effects. To be most useful, these sites should be coordinated with local meteorological and air quality observations and should enforce strict protocols for accuracy and cross-site calibrations.

Land-Use and Land-Cover Change

The mechanisms involved in land-atmosphere interactions are not well understood, let alone represented in climate models. A synergistic approach combining state-of-the-art models, field observations, and satellite imagery will be needed to advance our knowledge. Surface properties such as albedo, fractional vegetation coverage, emissivity, soil type, functional plant type, snow cover, and permafrost are examples of land-surface data that are needed. At the microscale, the use of very-high-resolution large-eddy simulations and micrometeorological observations from towers and low-flying, slow aircraft can elucidate some of the fundamental processes affecting the land-surface radiation balance through its interaction with turbulence and heat and momentum fluxes. Ever-increasing computing power now readily available allows very-high-resolution simulations with large-eddy simulations, including flow inside tree canopies.

At the mesoscale, land-cover heterogeneity triggers atmospheric circulations that enhance the heat and momentum fluxes in the atmospheric boundary layer and seem to increase the production of clouds. These circulations and the resulting cloud types and depths are sensitive to meteorological conditions and also depend on aerosol concentrations and size distribution. Satellite images have been key to identifying these types of clouds (Rabin and Martin, 1995). Yet there is still debate about the frequency of occurrence and intensity of clouds and precipitation resulting from such circulations, and their impact on the radiation balance (Weaver and Avissar, 2001; Doran and Zhong, 2002). Field campaigns at the mesoscale, which could be used to study these processes in more detail, are costly and complicated to perform. An integrated approach is needed involving a combination of satellite, aircraft, and tower observations. At the global scale, satellite imagery and models become even more important because in situ observations from ground stations and soundings are extremely limited. The challenge in modeling land-atmosphere interactions at that scale consists of including physical, chemical, and biological processes that occur at the microscale but propagate to the global scale through teleconnections. For example, global models have shown that the intensification of thunderstorm activity resulting from deforestation in Amazonia can affect precipitation in the U.S. Midwest. To capture these phenomena globally and with more accuracy, it is necessary to represent the global atmosphere at a very high resolution, which remains a challenge, even with the computing power available now. Appropriate parameterizations remain to be developed. Better datasets from more accurate and more frequent satellite observations are essential for the initialization and evaluation of global models.

Satellite MODIS data are promising in this regard because they can be used to globally monitor the land surface and its changes, seasonally and over longer time periods (Figure 6-3). The scientific value of MODIS is discussed in Running et al. (2004), Townshend and Justice (2002), and Schaaf et al. (2002). This instrumentation can be related to the longer-term measurements from the Advanced Very High Resolution Radiometer (AVHRR) and Landsat satellites as a monitor of land-use change and vegetation dynamics across several decades. Other satellite platforms to monitor land cover are reported in Bartalev et al. (2004) and include visible, infrared, and microwave wavelength sampling.

Field campaigns provide a complementary method to advance understanding of land surface processes. Campaigns such as BOREAS (for the boreal forest region of Manitoba and Saskatchewan), FIFE (for a 15 km by 15 km area in east-central Kansas), and others are summarized in Kabat et al. (2004). Such regionally specific programs permit the ground truthing of satellite data and provide higher spatial and temporal resolution.

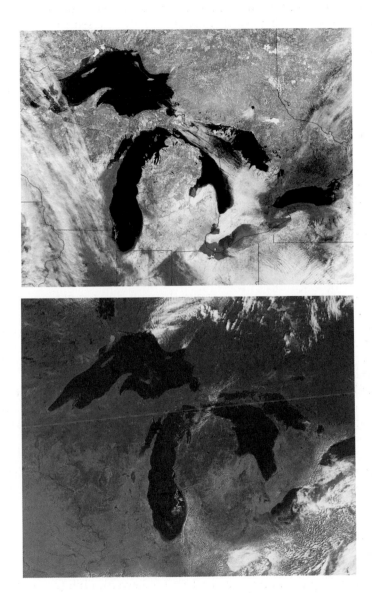

FIGURE 6-3 Observations of the land cover in the Great Lakes region of the United States and Canada, obtained by the Moderate Resolution Imaging Spectro-radiometer (MODIS) instrument aboard the Terra satellite. The top image was acquired on December 26, 2003, and the bottom image was acquired on September 16, 2002. SOURCE: NASA.

Solar

Space-based solar monitoring over the past 25 years covers more than two solar activity cycles and has established, unequivocally, the variability of the Sun's brightness at all wavelengths. Irradiance observations during the indefinite future are essential to quantify solar radiative forcing. Observations over only 2.5 cycles are insufficient to characterize the extremes of solar cycle irradiance variability or to detect speculated longer-term irradiance changes.

The extant record of total solar irradiance is compiled from observations made by half a dozen individual radiometers, cross-calibrated to account for individual absolute uncertainties and instabilities. Measuring solar irradiance with sufficient accuracy and repeatability to record true variability is a challenging radiometric task. The measurements must be made from absolute, electrically self-calibrating radiometers on space-based platforms. Significant drifts in instrument sensitivity can arise from changes in the space environment (solar exposure, thermal drifts, spacecraft pointing, power instabilities) and in optical and electrical components. These instabilities must be carefully removed from the radiometer signals.

Future observations of solar radiative forcing must address these challenges. Continuous, overlapping observations by multiple space-based radiometers are required. The overlap must be sufficiently long (of order one year or more) to provide the radiometric cross-calibration able to sustain the required long-term measurement accuracy. The cross-calibration must account for both the overall absolute level (typically traceable to uncertainties in the area of the primary aperture) and differences in temporal behavior (arising from instrumental drifts from many sources, but especially from solar exposure). The reliance of solar forcing observations on overlapping measurements will be alleviated only when absolute uncertainties are reduced by more than an order of magnitude relative to existing capability. Only then will benchmark measurements traceable to absolute standards be possible. Current expectations are that a new generation of phase-sensitive detection radiometers, the first of which is currently flying on the Solar Radiation and Climate Experiment (SORCE) mission, can achieve absolute accuracies of 0.01 percent. But this will have to be demonstrated by careful measurement validation and interpretation since the most recent SORCE measurements differ by 4 W m^{-2} (0.3 percent) from the historical database. This significant discrepancy motivates detailed scrutiny of all past and current radiometric observations so as to better identify and quantify sources of uncertainty in future measurements.

Future measurements of solar radiative forcing require not only measurements of total (spectrally integrated) irradiance but also simultaneous and self-consistent characterization of the spectrum of irradiance variability, which is strongly wavelength dependent. This is necessary because the multiple processes involved in solar radiative forcing are strongly wavelength dependent. SORCE observations capable of monitoring solar spectral irradiance from 0.2 to 2 μm have now begun. This continues the database of ultraviolet (UV) irradiance acquired by the Upper Atmosphere Research Satellite (UARS) since mid-1991 and commences a new, high-precision database of spectrally resolved irradiance observations, 25 years after the initiation of the total solar irradiance record.

The immediate challenge is to secure observations of both total and spectral irradiance that can connect the current measurements from SORCE (launched in 2002) with the eventual operational monitoring by the National Polar-orbiting Operational Environmental Satellite System (NPOESS). Current plans identify only total solar irradiance measurements during the intervening period. Once NPOESS has commenced observing solar radiative forcing, the challenge will be to obtain the necessary overlap of successive instruments. Current plans for the operation of the NPOESS spacecraft do not specify the needed (or any) overlap so that the long-term record may be jeopardized. A further issue is the probable lack of multiple observations of solar irradiance. Much has been learned in the past 25 years from comparison of independent radiometers at different stages of instrumental aging (and solar exposure). Multiple observations also provide crucial insurance against losing the entire long-term forcing record

Ocean Heat Content

Measurement of the ocean heat content provides an integrated method to monitor the radiative imbalance (Piexoto and Oort, 1992). This is because the ocean is the dominant heat storage location in the climate system (Levitus et al., 2000, 2001). The Argo network of ocean floats and satellite observations of ocean altimetry have been used to estimate trends in ocean heat content (Levitus et al., 2000, 2001; Willis et al., 2003). There is a need to better assess the frequency and the spatial coverage of ocean heat changes that are required to accurately determine the radiative imbalance on the order of tenths of a watt per square meter (e.g., Levitus et al., 2000, 2001). In addition, the data have to be evaluated in the context of radiative imbalance since 1995.

One possible approach for improving knowledge of past changes in oceanic heat content is to use ocean general circulation models (GCMs) to interpolate missing subsurface information in past decades from the increasingly sparse ocean measurements available back in time. Such ocean reanalysis projects are currently in their infancy (e.g., Carton et al., 2000). Another approach involves the use of sea level measurements to infer past changes in oceanic heat content. Such measurements are available both from modern satellite altimetry measurements, such as TOPEX/Poseidon data, that span roughly the past decade and sparser but longer-term tide gauge networks (Nerem and Mitchum, 2001). In order to infer ocean heat content changes from sea level estimates, however, one must make potentially restrictive assumptions regarding the thermal contributions to ocean heat content changes versus contributions from changes in continental run-off and glacial melting.

Metrology

Radiative forcings on climate have some overarching characteristics. They occur on intermediate to long timescales (longer than annual) that exceed the typical duration of the programs and systems that monitor them, and they involve the absorption, scattering, emission, and redistribution of electromagnetic radiation in the form of photons with a wide range of wavelengths and fluxes.

Because climate timescales are long, observations of radiative forcings and effects must be planned and maintained for the indefinite future. To secure real long-term changes and trends, the relevant geophysical quantities must be determined with a level of uncertainty (accuracy) that is significantly smaller than the expected changes. Thus far, the approach to achieve this relies on overlapping successive measurements to cross-calibrate their absolute uncertainties, which typically exceed the expected change. For example, total solar irradiance varies by about 0.1 percent from the minimum to the maximum of the 11-year solar cycle, but individual radiometers have absolute uncertainties of only 0.2 percent. A composite irradiance record is thus possible only by cross-calibrating individual radiometers (Fröhlich and Lean, 2002), taking into account as well the effect of sensitivity drifts and environmental (e.g., space platform) influences on their long-term repeatabilities (also called long-term precision or relative accuracy). This situation is equally true for the tropospheric temperature record constructed from observations made by Microwave Sounding Units (NRC, 2000). Questions remaining about the reliability of these long-term composite records undermine the certainty with which the parameters are known.

A secure long-term database of radiative forcings and effects requires that the accuracies of the geophysical parameters ultimately be tied to irrefutable absolute standards that are tested and validated in perpetuity for uncertainty and repeatability. Also essential is a requirement that total forcing be documented along with the forcing due to individual components. Anderson et al. (2003a) point out that the uncertainty in net forcing is much greater than the forcing due to an individual forcing agent. Benchmark measurements of radiative forcings and climate parameters are needed immediately to provide records of absolute values for all time of a number of carefully selected observables that define climate forcings and climate responses. Since radiative forcings and climate responses are highly wavelength dependent, high spectral resolution is needed to isolate the spectral signatures of the relevant processes and components. This produces additional challenges since accuracy and calibration difficulty increases as spectral resolution increases (stray light, instrument profile function, wavelength calibration, signal to noise, matching to available standards). Key parameters for which benchmark measurements are crucial include among others sea level altimetry, solar irradiance, global positioning system (GPS) index of refraction, ozone and CO_2 concentrations, and spectrally resolved, absolute radiance to space.

Establishing and validating the accuracy and precision of a geophysical quantity involves tracking instrument calibrations from the laboratory to deployment and throughout mission lifetimes to reduce systematic errors on orbit. This requires considerable additional effort and commitment to an experimental strategy designed to reveal systematic errors and drifts through independent cross checks, open inspection, and continuous interrogation. It involves simultaneous observations of related and similar quantities using both similar and differing radiometric techniques. Regular calibrations are needed, for example, using the Sun, Moon, known land scenes, or on-orbit sources or detectors. Since the forcings and responses that determine any one particular climate state involve a distribution about a mean, the ensemble must be properly characterized and quantified so that changes in the mean can be identified reliably. Ultimately, the specification of the forcings and responses must be integrated to test climate forecast models.

Achieving radiometric accuracy and traceability requires new programs and techniques to advance the current state of metrology and transfer these advances to the determination of radiative climate forcings and effects. This has motivated an alliance of NPOESS and the National Institute of Standards and Technology (NIST), but there is a need to accord this a high priority with sufficient time and funds. It is unlikely that many of the climate variables measured by NPOESS will be directly traceable to an absolute standard. A recent NIST workshop (Ohring et al., 2004) to address this challenge notes that "measuring the small changes associated with long-term global climate change from space is a daunting task. Satellite instruments must be capable of observing atmospheric temperature trends as small as 0.1°C per decade, ozone changes of as little as 1 percent per decade and variations in the Sun's output as tiny as 0.1 percent per decade." NIST is developing new facilities to meet the metrology challenges of future climate-related observations. Of particular relevance is the Spectral Irradiance and Radiance Calibrations with Uniform Sources (SIRCUS), which has the ability to characterize and calibrate radiometric instruments with spectrally pure sources of differing flux levels over a wide range of wavelengths (ultraviolet to infrared), with detectors tied to cryogenic cavity radiometers (absolute accuracy 0.001 percent for power measurement) and apertures measured at NIST.

LABORATORY AND PROCESS STUDIES FOR AEROSOLS

Key uncertainties in the composition and properties of aerosol particles and their role in clouds require measurements and models under controlled conditions. Laboratory and process studies are needed to resolve these questions by four types of measurements. First, measurement of the composition and thermodynamic properties of organic and inorganic components, together with development of intelligent parameterizations of these properties, is necessary to describe the properties of individual particles. Second, measurements of spectrally resolved imaginary refractive indices are needed to determine absorbing properties. The third type of measurement characterizes the morphology and reactivity of particle surfaces. Finally, surface tension and wettability of organic particles must be measured in order to predict cloud droplet activation properties.

The thermodynamic properties of organic and some mineral components are not well understood. Most particle types show a strong hysteresis effect between the relative humidity for deliquescence (conversion from solid to liquid) and the relative humidity for efflorescence (conversion from liquid to solid). This hysteresis effect is not well characterized in the laboratory or in the field, yet it plays a critical role in particle optical properties. The partitioning of semivolatile organic compounds between the gas and the aerosol phase also needs to be better determined as a function of temperature and aerosol phase composition.

The optical properties of organic compounds present in aerosols are poorly known. Much of the existing information is limited to the UV and driven by the needs of the polymer industry. Very little information exists for spectrally resolved imaginary refractive indices in the visible spectrum. Because these measurements are relatively routine but very time-consuming, there has been little interest in the research community in collecting the required database of optical properties.

The third type of measurement is the reactivity of ambient particles, given their shapes and structures. Such reactivity may include probabilities of inorganic and organic reactions that affect particle lifetime, distribution, and optical behavior. Porosity and surface area strongly determine the rates and yields of heterogeneous chemical reactions, yet very little is known about these characteristics for ambient particles.

A fourth property of chemical mixtures that is not well understood for pure or mixed organic or inorganic particles is their surface structure and wetting behavior. Surface tension plays an important role in the indirect effect of aerosol particles, potentially providing an important determining factor for the particles to activate. Organic compounds in particles may significantly alter the efficiency with which particles can serve as cloud condensation nuclei (Facchini et al., 2000; Feingold and Chuang, 2002; Ming and Russell, 2004). The transformation timescale from hydrophobic to hydrophilic states is a seriously uncertain parameter in current models.

ATMOSPHERIC REANALYSIS AND DATA ASSIMILATION

Atmospheric reanalysis involves using models to interpolate observations in order to construct physically consistent estimates of atmospheric structure and dynamics. The National Centers for Environmental Prediction (NCEP) Reanalysis (Kalnay et al., 1996) and the European Centre for Medium-Range Weather Forecasts (ECMWF) Reanalysis (ERA-40) (Bengtsson et al., 2004) are two global analyses that extend across several decades and will continue into the future. Reanalyses can be used to assess the change over time of selected space- and time-integrated climate metrics, such as the 1000-500 mb thickness, the 200 mb heights, tropopause height, and the 200 mb winds (Chase et al., 2000b; Pielke et al., 2001; Santer et al., 2003b).

It remains difficult, however, to estimate reliable, small-amplitude trends from reanalyses (Bengtsson et al., 2004), owing mainly to temporal variations in input data quantity and quality. Given these heterogeneities in reanalyses, it is essential to determine the magnitude of trends that must occur before they can be determined to be statistically significant (Chase et al., 2000b). The use of metrics that integrate atmospheric structure and dynamics represents another effective procedure to utilize reanalyses for trend assessments in that the effect of heterogeneities in the data record may be reduced. Examples include the thickness between pressure surfaces, tropopause height, or the vertical wind shear across the troposphere. The first two provide vertically integrated measures of the warming of the troposphere in response to radiative heating. The third provides an integrated measure of the horizontal gradient in tropospheric mean temperature.

Future reanalyses should strive for as homogeneous a dataset as possible to monitor temporal and spatial changes in tropospheric heat content. This information would be valuable in relating to observed temporal and spatial changes in ocean heat content. For example, can the atmospheric reanalyses help explain the observed focusing of ocean warming in the midlatitudes of the Southern Hemisphere, and will this continue into the future? Accurate reanalyses can also address the question of whether the difference between surface and tropospheric temperature trends is real or a product of inconsistencies in monitoring.

RELATING CONCENTRATIONS OF GREENHOUSE GASES AND AEROSOLS TO SOURCES

An important step in understanding human and natural impacts on climate is relating what is known about sources of greenhouse gases and aerosols to their observed abundances in the atmosphere. Understanding this link is especially challenging for those atmospheric species that are produced in the atmosphere by chemical reactions of precursor species, have short atmospheric lifetimes, or have a multitude of sources. Two modeling tools—chemical transport models (CTMs) and inverse models—have been developed to assist scientists in relating sources to atmospheric concentrations.

Chemical Transport Model Analyses

Aerosols and ozone have short atmospheric lifetimes and hence inhomogeneous atmospheric distributions. Radiative forcing calculations for these species require global three-dimensional characterization of their concentration fields, the evolution of these concentration fields with time, and correlations with other radiative forcing agents such as clouds and water vapor. This is generally done with CTMs that solve the continuity equation for the species of interest using information on sources, transport, chemical processes, and deposition. CTM simulations provide the basis for the current Intergovernmental Panel on Climate Change (IPCC, 2001) estimates of the radiative forcings from aerosols and tropospheric ozone. They need to be improved in the future by assimilating high-density chemical observations from satellites, using algorithms similar to those presently implemented for meteorological data assimilation. This is already done routinely for stratospheric ozone (Stajner et al., 2001) and should be extended to satellite observations of tropospheric ozone and its precursors (including nitrogen dioxide [NO_2], formaldehyde [HCHO], and carbon monoxide [CO]), aerosol optical depths, and aerosol size distributions (Figure 6-4). Eventually, chemical data assimilation and the associated CTM calculations should be done within GCMs and coupled with meteorological data assimilation. This approach will have the advantage of better accounting for correlations with clouds and water vapor. It will also resolve the synoptic-scale coupling of the radiative effects and the meteorological response, as well as coupling interactions between aerosol and cloud processes (Koch et al., 1999; Mickley et al., 1999).

Several elements of stratospheric forcings from changes in ozone and

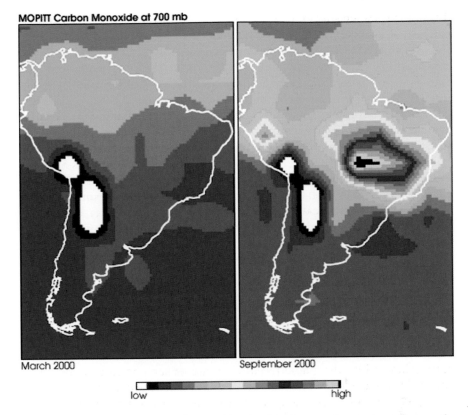

MOPITT Carbon Monoxide at 700 mb

March 2000   September 2000

low   high

FIGURE 6-4 Carbon monoxide at 700 mb altitude over South America, as observed by the Measurements Of Pollution In The Troposphere (MOPITT) sensor flying aboard NASA's Terra spacecraft, and assimilated into an atmospheric chemical transport model using wind vectors provided by the National Center for Environmental Prediction (NCEP). Data for producing the image on the left were acquired on March 3, 2000, when fairly low levels of CO were observed, and for the image on the right on September 7, 2000, when a large carbon monoxide plume is seen over Brazil. The generally higher carbon monoxide levels in September are attributed to South American fire emissions and the transport of carbon monoxide across the Atlantic Ocean from Southern Africa fires. SOURCE: NASA Goddard Space Flight Center.

volcanic aerosols are now very well simulated. In the case of stratospheric ozone, the resulting stratospheric cooling is an integral component of the forcing, and the simulated temperature changes match reasonably well with observations (Ramaswamy and Schwarzkopf, 2002; Schwarzkopf and Ramaswamy, 2002; Shine et al., 2003). In the case of volcanic aerosols, models have performed useful comparison exercises (e.g., Pollack et al., 1993). The 1991 Mt. Pinatubo eruption has provided a number of tests against which model simulations can be verified. Stratospheric warming observed after Pinatubo is well simulated by models that employ the detailed spatial-temporal evolution of the particles and incorporate them in a multiwavelength radiative transfer code within a reasonable GCM (Ramaswamy et al., 2004). Indeed, the warming resulting from this eruption, the radiative flux comparisons with satellite observations, the cooling of the troposphere, the change in precipitable water, and the winter warming in northern high latitudes are all at least qualitatively well simulated, attesting to a degree of confidence in the working of climate models (Ramachandran et al., 2000; Ramaswamy et al., 2004; Soden et al., 2002; Stenchikov et al., 2002).

Global CTM simulations of stratospheric and tropospheric ozone are now fairly mature (IPCC, 2001). However, great difficulties remain in the simulation of transport across the tropopause, where ozone has its largest radiative effect. Most CTMs have excessive cross-tropopause transport of air (Tan et al., 2004), at least in part because of noise in the vertical winds induced by the meteorological data assimilation process. In addition, CTMs tend to greatly underestimate the observed trend of tropospheric ozone over the past century (Mickley et al., 2001; Shindell and Faluvegi, 2002) and over the past decades (Fusco and Logan, 2003), suggesting some fundamental difficulty in relating tropospheric ozone concentrations to their sources. Addressing this issue will require focused studies of regional-scale budgets of ozone and its precursors, as well as improved understanding of the natural sources of tropospheric ozone precursors including fires, lightning, and vegetation.

Global CTM studies of aerosols are still in their infancy. Sources of radiatively important aerosol types including organic carbon, elemental carbon, dust, and sea salt are highly uncertain and crudely parameterized. There are relatively good constraints on emissions of sulfur gases, but oxidation to form sulfate aerosols takes place principally in clouds and is thus strongly tied to the simulation of the hydrological cycle (which is highly uncertain). Loss of aerosols occurs mainly by wet deposition, which is subgrid scale for global models and thus has to be parameterized. Better coupling of aerosols with the hydrological cycle is needed; joint data assimilation of aerosol, cloud, and precipitation properties should be pursued in the future. However, assimilation techniques also have fundamental limitations (e.g., lack of knowledge on subgrid scales, inadequate diagnoses of vertical velocities, possible inconsistency between reality and assimilation model physics) that could have a significant impact, especially on the concentrations of short-lived species.

Almost all global CTM studies of aerosols so far have been mass-only simulations that do not resolve the aerosol size distribution, mixing across components, or phase. This is evidently problematic for radiative forcing calculations and, in particular, prevents simulations of the indirect effect except through loose empirical relationships between cloud droplet number concentrations and preexisting aerosol mass concentrations (Boucher and Lohmann, 1995). There is a major computational problem because accounting for aerosol microphysics and allowing for an ensemble of aerosol mixing states rapidly increases the number of prognostic model variables. It appears unlikely that this problem will be solved over the next decade by simple increases in computing resources. Innovative algorithms for simulating aerosol microphysics are needed, such as the method of moments (McGraw, 1997) or new sectional approaches (Adams et al., 2003). Better understanding is also needed of the fundamental processes driving aerosol microphysics, particularly nucleation.

Inverse Models

The standard way for specifying emission inventories in CTMs uses "bottom-up" approaches in which knowledge of the underlying processes, and of the associated emission factors, is parameterized and extrapolated on the basis of globally available socioeconomic or environmental information. The bottom-up approach provides the fundamental tool for ascribing sources to specific emission processes and for making future projections. However, there are often large uncertainties in the emission factors and their extrapolation. One can attempt to reduce this uncertainty with "top-down" constraints on emissions that combine information on observed atmospheric concentrations with CTM-derived relationships between concentrations and sources. Formal inverse models combine these bottom-up and top-down approaches by seeking an optimum solution for the emissions that best accommodates the a priori constraints from bottom-up inventories and information from observations (Kasibhatla et al., 2002).

Global observations from long-term surface-based networks (e.g., NOAA CMDL and ALE/GAGE networks) have been used extensively in inverse model studies of sources for CO_2 (e.g., Peylin et al., 2002), CO (e.g. Kasibhatla et al., 2002; Petron et al., 2002), methane (Wang et al., 2004), and halocarbons (Mahowald et al., 1997). Inverse model studies for CO_2 have played a key role in quantifying the terrestrial sink of CO_2 at northern midlatitudes. Observations from aircraft campaigns and from satellites are presently increasing the scope and possibilities of inverse methods (Arellano et al., 2004; Palmer et al., 2003; Heald et al., 2004). Variational data assimilation methods are now being developed to improve the detail in the characterization of sources enabled by large observational datasets (e.g., Kaminski et al., 2002). Future inverse model studies should make use of available observations of aerosol surface concentrations and optical depths, as well as the information contained in the observed correlations between species concentrations, for example, between CO_2 and CO (Suntharalingam et al., 2004) or methane and ethane (Xiao et al., 2004). These correlations can improve the top-down constraints on the sources and also reduce the errors associated with CTM transport.

CLIMATE FORCING AND RESPONSE OVER EARTH'S HISTORY

A comprehensive database of radiative forcings and effects exists primarily for the past 25 years because many of the relevant observations require space-based observations. Present in this epoch are two major volcanic eruptions (El Chichon and Mt. Pinatubo), a few significant El Niños (1983, 1997), and two solar irradiance cycles. The reconstruction of much longer-term records of forcings and effects is crucial for a broader perspective.

Empirical analyses of correlations between adopted radiative forcing histories and climate reconstructions provide exploratory but limited insights into the relative roles of radiative forcings of climate change in the recent past (e.g., Lean et al., 1995; Mann et al., 1998; Waple et al., 2002). Correlations of various proxies of climate change and radiative forcings during the Holocene suggest the influences of solar variability and orbital motions on a range of climate phenomena including drought (Hodell et al., 2001), rainfall (Neff et al., 2001), and North Atlantic winds and surface hydrography (Bond et al., 2001). Other studies characterize the evolution of variability modes as sources of historical climate change, including the Arctic Oscillation (Noren et al., 2002) and the El Niño/Southern Oscillation (ENSO; Moy et al., 2002). Another type of forcing response investigation is the effect of ice sheet changes during the last glacial maximum (e.g., Manabe and Broccoli, 1985).

Detailed physical insight into the role of past natural radiative forcing requires that documented climate reconstructions be compared with model simulations driven by the actual geophysical forcings. However, some current limitations hamper our ability to draw precise conclusions from such comparisons, even in the recent past. Moderate differences exist, for example, between various alternative reconstructions of past hemispheric temperature trends (e.g., Folland et al., 2001; Jones et al., 2001; Mann et al., 2003; see Jones and Mann, 2004, for a comparison of multiple reconstructions). A reduction of uncertainties in these reconstructions, along with a resolution of differences among competing estimates, is essential to improve knowledge of the precise history of large-scale mean temperature changes in past centuries, and hence of radiative forcing effects. Such a resolution is likely to come from the availability of increased high-quality proxy reconstructions in key regions, particularly in the data-sparse regions of the tropical oceans and Southern Hemisphere. Improved specification of physical differences and limitations of various temperature proxies (tree rings versus boreholes versus corals) is also needed.

There is a broadly consistent view between different climate models and empirical proxy-based reconstructions of hemispheric mean surface temperature changes in past centuries. The models indicate that greenhouse gases explain the observed 0.6°C global surface warming in the past three decades and that some combination of solar and volcanic forcings is likely responsible for temperature fluctuations of a few tenths of a degree Celsius in the preindustrial period (IPCC, 2001). Model and observational studies suggest that land-cover change may account for some of the surface temperature variation over land (e.g., Kalnay and Cai, 2003; Marshall et al., 2003).

However, there are also significant differences among the model simulations. These differences arise from a number of sources (see Jones and Mann, 2004), including (1) differences in the sensitivities of the models to radiative forcing, which vary by as much as a factor of two; (2) differences in the reconstructed radiative forcings used to drive the model simulations; and (3) differences in the way that radiative forcing estimates are represented in the model. For example, in the case of volcanic aerosols, some models impose a fixed annual mean TOA radiative forcing simply by changing the solar constant (Gonzalez-Rouco et al., 2003), while others (e.g., Shindell et al., 2003, 2004) specify the forcing on a seasonally, latitudinally, and vertically resolved basis. It is clear that improved estimates of past radiative forcing changes and a more organized community-wide effort to perform a controlled set of simulations using common forcing estimates could help to resolve these differences.

Spatial patterns of climate change are difficult to compare between models and observations. The dearth of proxy data over large parts of the oceans in past centuries restricts the spatial detail available in current proxy-based reconstructions (Jones and Mann, 2004). Moreover, at regional spatial scales, the role of internal, unforced variability in the climate (which is intrinsically irreproducible by a forced simulation) is likely to be greater, and observed variations may be dominated by influences from large-scale modes of atmospheric circulation such as the North Atlantic Oscillation (NAO) and ENSO. Although there has been some success in reproducing past reconstructed changes in model simulations, including an NAO-like response to radiative forcing changes, experiments employing fully coupled land-ocean-atmosphere models to study regional past climate change are just now under way. It is likely that details of stratospheric dynamics and chemistry, ocean circulation, vegetation and soil dynamics, and mechanisms of land-ocean-atmosphere coupling are all important in describing past regional-scale changes in climate. A particular challenge is to quantify the role of radiative forcings (versus other mechanisms) in effecting coherent climate change in widely separated geographical regions, as is evident in paleoclimate proxies on multiple and often abrupt timescales (Rial et al., 2004).

CLIMATE MODELS

Applications of climate models include developing better understanding of processes and predicting future conditions. Compared to simulating the weather, climate modeling faces the challenges of longer timescales, ranging from years to centuries and longer. Climate modeling also requires the accurate simulation of each important component of the climate system, including the atmosphere, oceans, land surface, and continental ice fields, as well as realistic estimates of external forcings (i.e., solar, volcanoes). Physical, biological, and chemical processes taking place in each of these components interact with each other across the spectrum of space and timescales. In simulating future climate, models must take into account how humans will affect emissions of greenhouse gases and aerosols as well as modify land use and land cover. Because future human activities are inherently uncertain, model projections of future climate are typically computed for multiple scenarios of future emissions.

Historical data have been used extensively to evaluate climate models. The Atmospheric Model Intercomparison Project (AMIP) is an excellent example of model validation (Gates et al., 1998) based on archived atmosphere and sea surface data. Such model evaluations need to be extended to encompass the spectrum of important climate forcing effects on such societally important quantities as water resources, agricultural and natural vegetation growth, and air pollution. Can skillful forecasts of changes in these quantities be made as a function of radiative and other climate forcings? These issues are regional in scale, such that validation of model process simulation and forecast skill must be completed at these subglobal scales.

A particular challenge for global climate models is modeling forced climate change over the last few decades. This is the time period with the greatest change in well-mixed greenhouse gases as well as the most complete observational datasets. Some studies have found discrepancies between the surface and tropospheric surface temperature changes in simulations and observations (Chase et al., 2004), which could be attributed to deficiencies in either models or observations, or a combination (NRC, 2001; Christy and Norris, 2004; Mears et al., 2003; Vinnokov and Grody, 2003, Pielke and Chase, 2004; Fu et al., 2004). Other studies, however, find good agreement between observations and the model-predicted spatial and vertical fingerprints of radiatively forced climate change in recent decades (Allen et al., 2000; Stott et al., 2000; Wigley et al., 2000; Barnett et al., 2001; Santer et al., 2000, 2003a,b; Karoly et al., 2003). Additional evaluations of the ability of models to reproduce regional and global climate in recent decades—including tropospheric temperature, ocean heat content, and other climate variables in addition to surface temperature—should be a major priority for further quantifying model predictive skill. Models should also be encouraged to incorporate forward radiance calculations as model diagnostics to compare with observed radiances.

In order to narrow down the uncertainties associated with radiative forcing effects on climate, models have to be improved in many aspects. Of particular importance is improving the representation of cloud processes, the coupling between the atmosphere and the land surface and ocean, the impacts of regional variability in diabatic heating, and the simulation of regional-scale climate.

Clouds and Microphysics

Uncertainties in relating aerosol to cloud droplet populations seriously limits our ability to quantify the indirect aerosol effects. To treat cloud droplet formation accurately, the aerosol number concentration, its chemical composition, and the vertical velocity on the cloud scale need to be known. Abdul-Razzak and Ghan (2000) developed a parameterization based on the Köhler theory that can describe cloud droplet formation for a multimodal aerosol. This approach has been extended by Nenes and Seinfeld (2003) to include kinetic effects, that is, considering that the largest aerosols do not have time to grow to their equilibrium size. To apply one of these parameterizations, the updraft velocity relevant for cloud formation needs to be known. Some climate models apply a Gaussian distribution or use the turbulent kinetic energy as a surrogate for updraft velocity (Ghan et al., 1997; Lohmann et al., 1999). Others avoid this issue completely and use empirical relationships between aerosol mass and cloud droplet number concentration instead (Menon et al., 2002a). This method is limited because of the scarce observational database. At present, the relationship can only be derived between cloud droplet number and sulfate aerosols, sea salt, and organic carbon; no concurrent data for dust or black carbon and cloud droplet number are available yet. Therefore, and because of their universality, the physically based approaches described formerly should be used in future studies of aerosol-cloud interactions.

Since the first IPCC assessment, great improvements have been made in the description of cloud microphysics for large-scale clouds. Whereas early studies diagnosed cloud amount based on relative humidity, most models now predict cloud condensate in large-scale clouds. The degree of sophistication varies from just predicting the sum of cloud water and ice (Rasch and Kristjánsson, 1998) to predicting cloud water, cloud ice, snow, and rain as separate species (Fowler et al., 1996). Because the aerosol indirect effect is based on the change in cloud droplet number concentration, some models predict cloud droplet number concentrations using one of the above-described physically based aerosol activation schemes as a source term for cloud droplets (Ghan et al., 1997; Lohmann et al., 1999). There is currently a great discrepancy in models between the sophisticated treatment of cloud microphysics in large-scale clouds and their very rudimentary treatment in convective clouds. Furthermore, there is a mismatch between aerosol activation and cloud formation in most climate models because cloud formation relies on a saturation adjustment scheme whereas aerosol activation relies on a subgrid-scale vertical velocity. Part of this problem will be solved within the next decade when climate models can be run at higher spatial resolution and with smaller time steps.

Including Land Surface Models

Changes in land use pose a nonnegligible climate forcing as well. Climate models are just beginning to include detailed land surface models that are coupled to the simulation of the atmosphere. Also, carbon-cycle feedbacks have been shown to be very important in predicting climate change over the next century (e.g., Schimel et al., 2001; Jones et al., 2003). One important question is whether the terrestrial carbon cycle becomes a net source of carbon dioxide during the next century. To address this issue, vegetation-meteorology-biogeochemical cycle interactions need to be included in climate models.

Diabatic Forcing Heterogeneity

A variety of heterogeneous diabatic forcings have been shown to alter the climate both in the region where this forcing occurs and at large distances through teleconnections. These forcings include land-cover change and vegetation dynamics, soil moisture, ocean color, and aerosols (e.g., Chung and Ramanathan, 2003; Shell et al., 2003; Claussen et al., 2004). On the regional scale, there is general agreement on the importance of these regional forcings on climate as summarized by Kabat et al. (2004). However, despite the plausible scientific basis as to why teleconnections should be expected and the analog to ENSO events, the global teleconnections associated with these regional forcings are not as widely accepted. The argument against the robustness of the long-range connectivity involves possible oversensitivity of the climate models that have been used in the studies and the statistical significance of the results.

To address these comments, climate models with appropriate sensitivity and resolution should be used to perform experiments with observed regional anomalies of diabatic forcing, as well as with realistic perturbation simulations (such as between natural and current landscapes). The results should be tested statistically to assess the robustness of any differences. Van den Hurk et al. (2003), for example, conducted three ensembles of five runs each: the control ensemble used constant global leaf area index (LAI) values; the second ensemble used seasonally varying LAI fields; and a third ensemble used the same seasonally varying LAI fields but with a noise term added. This methodology should be adopted for each of the regional diabatic forcings. Sufficient computer resources are required for these computationally expensive integrations.

Simulating Regional Climate

A summary of the current state of regional climate modeling is reported in Kabat et al. (2004). A major new direction is the dynamic coupling between the regional atmosphere and land surface and between the atmosphere and oceans (e.g., Eastman et al., 2001a,b). Coupled atmosphere-sea ice simulations are also being performed. The incorporation of atmospheric chemistry, including aerosol effects, also needs to be included in this dynamic coupling. Matsui et al. (2004), for example, show the sensitivity of the aerosol effects on cloud and precipitation processes due to environmental thermodynamic structure. These modeling tools will permit the investigation of the role of regional radiative forcing in altering regional climate as well as high-spatial-resolution estimates of the ability of regional climate change and variability to teleconnect to other regions and globally.

7

Recommendations

The current global mean top-of-the-atmosphere (TOA) radiative forcing concept with adjusted stratospheric temperatures has been used extensively in the climate research literature over the past few decades and has also become a standard tool for policy analysis endorsed by the Intergovernmental Panel on Climate Change (IPCC). It is a useful index for estimating global average surface temperature change resulting from changes in well-mixed greenhouse gases, solar irradiance, surface albedo, and nonabsorbing aerosols. The relative ease of calculating radiative forcing and the associated temperature response has enabled the use of climate models, simpler versions of those models, and chemical transport models to investigate the many factors that may influence climate. In short, the TOA radiative forcing concept still has considerable value and should be retained as a standard metric in future climate research.

Nonetheless, the traditional radiative forcing concept has major limitations that have been revealed by recent research on nonconventional forcing agents and regional studies. It is limited in its ability to describe the climate effects of absorbing aerosols, aerosol interactions with clouds, ozone, land-surface modification, and surface biogeochemical effects. Also, it diagnoses only one measure of climate change: equilibrium response of global mean surface temperature. It does not provide information on nonradiative climate effects, spatial or temporal variation of the forcing, or nonlinearity in the relationship between forcings and surface temperature response. Recent extensions of the concept that allow surface temperatures to adjust have refined the radiative forcing concept to address deficiencies in the original approach. Although currently applied to global mean conditions, this method could be extended for regional conditions.

The strengths of the traditional radiative forcing concept warrant its continued use in scientific investigations, climate change assessments, and policy applications. At the same time, its limitations call for using additional metrics that account more fully for the nonradiative effects of forcing, the spatial and temporal heterogeneity of forcing, and nonlinearities. The committee believes that these limitations can be addressed effectively through the introduction of additional forcing metrics in climate change research and policy. This chapter provides several recommendations for extending the traditional radiative forcing concept in the scientific and policy arenas. It identifies research needed to improve quantification and understanding of different forcings and their impacts on climate, to better inform climate policy discussions, and to obtain reliable observations of climate forcings and responses in the past and future. A large number of recommendations are provided because many research avenues need to be explored in order to improve understanding of climate forcings. The recommendations that should be undertaken immediately with high priority are identified with the ❖ symbol.

EXPANDING THE RADIATIVE FORCING CONCEPT

Account for the Vertical Structure of Radiative Forcing

Recent observations show that radiative forcing calculated at the top of the atmosphere is not always a good index for changes in surface temperature. Indeed, the relationship between TOA radiative forcing and surface temperature is not valid if there is significant variation in the vertical distribution of radiative forcing. For example, the direct radiative forcing of black carbon and other absorbing aerosols leads to a reduction in surface heat input while increasing atmospheric heating. Likewise, land-use changes can modify latent and sensible heat fluxes at the surface. Considering the surface radiative forcing along with the comparable value at the top of the atmosphere would enable quantification of the effects of aerosols and other forcings on the surface energy balance and the net forcing of the atmosphere. It would provide information about the extent to which forcings affect the atmospheric lapse rate, with implications for precipitation and mixing.

Associated with expanding the treatment of radiative forcing in this way are several new research needs. In general, climate models have been unable to reproduce the vertical distribution of forcing due to aerosols observed during aircraft campaigns. Nor, in general, do general circulation models (GCMs) have the needed stratospheric processes to adequately model volcanic and solar ultraviolet radiation effects. Little research has addressed how climate response might depend on the vertical structure of the radiative forcing.

RECOMMENDATIONS:

❖ Test and improve the ability of climate models to reproduce the observed vertical structure of forcing for a variety of locations and forcing conditions.

❖ Undertake research to characterize the dependence of climate response on the vertical structure of radiative forcing.

❖ Report global mean radiative forcing at *both* the surface and the top of the atmosphere in climate change assessments.

• Develop parameterizations for using surface forcing in integrated assessment and simple climate models.

Determine the Importance of Regional Variation in Radiative Forcing

The concept of a global mean radiative forcing is an approximation. Even forcings thought to be fairly uniform, such as solar variability and the well-mixed greenhouse gases, have seasonal and latitudinal variability. Other forcings, in particular tropospheric aerosols and landscape changes, have much more spatial and temporal heterogeneity in their distribution. Human modifications to landscape and vegetation dynamics have caused large regional changes in the surface distribution of net absorbed surface radiation into latent and sensible turbulent heat fluxes. To date, there are only limited observational and modeling studies of regional radiative forcing and response. Indeed, there is not yet a consensus on how best to diagnose a regional forcing and response in the observational record.

Regional variations in radiative forcing are likely important for understanding regional and global climate responses; however, the relationship between the two is not well understood. Regional climate responses can also be caused by global forcings, making it difficult to disentangle the effects of regional and global forcings. Regional diabatic heating can cause nonlinear, long-distance communication of convergence and divergence fields, often referred to as teleconnections. Thus, regionally concentrated diabatic heating can influence climate thousands of kilometers away from its source region. Improving societally relevant projections of regional impacts will require a better understanding of the magnitude of regional forcings and the associated climate response.

RECOMMENDATIONS:

❖ Use climate records to investigate relationships between regional radiative forcing (e.g., land-use or aerosol changes) and climate response in the same region, other regions, and globally.

• Test and improve model simulations of regional radiative forcing and the surface energy budget using observations from aircraft campaigns, surface networks, and satellites.

❖ Quantify and compare climate responses from regional radiative forcings in different climate models and on different timescales (e.g., seasonal, interannual), and report results in climate change assessments. Specific focus should be given to

— regions in which forcing could interact with modes of climate variability (e.g., El Niño/Southern Oscillation [ENSO], Antarctic Oscillation, Arctic Oscillation) and result in major teleconnections (e.g., forcing over the tropical Pacific from biomass burning affecting ENSO and therefore drought in the United States, Australia, and other distant regions);
— regions in which there are significant anthropogenic forcings due to anthropogenic emissions or land-use modifications—for example, North America and Europe (industrial emissions, reduction in sulfur dioxide [SO_2] emissions), Asia (black carbon emissions, land-use change), and the Amazon (deforestation); and
— major geopolitical regions with large anticipated socioeconomic changes or vulnerability to climate change and variability.

Determine the Importance of Nonradiative Forcings

Several types of forcings—most notably aerosols, land-use and land-cover change, and modifications to biogeochemistry—impact the climate system in nonradiative ways, in particular by modifying the hydrological cycle and vegetation dynamics. Aerosols exert a forcing on the hydrological cycle by modifying cloud condensation nuclei, ice nuclei, precipitation efficiency, and the ratio between solar direct and diffuse radiation received. These aerosol forcings are sometimes referred to as thermodynamic forcings because they affect spatial patterns of diabatic heating. In some cases, aerosols may be able to modify the hydrological cycle without changing the global average surface temperature. Other nonradiative forcings modify the biological components of the climate system by changing the fluxes of trace gases and heat between vegetation, soils, and the atmosphere; the biogeochemistry of vegetation biomass and soils; or plant species composition. Nonradiative forcings have been shown in a few studies to have first-order effects on regional and global climate, although the globally averaged impacts are not yet sufficiently quantified to allow a careful comparison with forcing from greenhouse gases.

No metrics for quantifying nonradiative forcing have been accepted. Unlike traditional radiative forcing, which can be directly related to surface temperature, nonradiative forcings are not easily linked to a single climate variable. No single metric will be applicable to all nonradiative forcings. Nonradiative forcings generally do have radiative impacts, so one option would be to compare them by quantifying these radiative impacts. Although this approach would enable comparisons with traditional radiative forcings, it would not convey fully the impacts of nonradiative forcings on societally relevant climate variables, such as precipitation or ecosystem functioning. Furthermore, quantifying nonradiative forcings in terms of their radiative effects is not straightforward. Another consideration in identifying potential metrics for nonradiative forcings is their significant regional variation; any new metrics will have to be able to characterize the regional structure in forcing and climate response. Further work is needed to quantify links between regional nonradiative forcing and climate response, whether the response occurs in the region, in a distant region through teleconnections, or globally.

RECOMMENDATIONS:
❖   Improve understanding and parameterizations of aerosol-cloud thermodynamic interactions and land-atmosphere interactions in climate models in order to quantify the impacts of these nonradiative forcings on both regional and global scales.
❖   Develop improved land-use and land-cover classifications at high resolution for the past and present, as well as scenarios for the future.
•   Develop parameterizations of terrestrial and marine biogeochemistry to investigate the resulting nonradiative forcings.
•   Identify suitable climate diagnostics, metrics, and monitoring procedures for specific nonradiative forcing processes and responses.

ADDRESSING KEY UNCERTAINTIES

Whereas the level of understanding associated with radiative forcing by well-mixed greenhouse gases is relatively high, there are major gaps in understanding for the other forcings, as well as for the links between forcings and climate response. Error bars remain large for current estimates of radiative forcing by ozone, and are even larger for estimates of radiative forcing by aerosols. Nonradiative forcings are even less well understood. The potential for large and abrupt climate change triggered by radiative and nonradiative forcings needs to be explored. The following recommendations identify critical research avenues for addressing these key uncertainties.

Reduce Uncertainties Associated with Indirect Aerosol Radiative Forcing

The interactions between aerosols and clouds can lead to a number of indirect radiative effects, which arguably represent the largest uncertainty in current radiative forcing assessments. In the so-called first indirect aerosol effect, the presence of aerosols leads to clouds with more, but smaller particles; such clouds are more reflective and therefore have a negative radiative forcing. These smaller cloud droplets can also decrease the precipitation efficiency and prolong cloud lifetime; this is known as the second indirect aerosol effect. The so-called semidirect aerosol effect occurs when absorption of solar radiation by soot leads to an evaporation of cloud droplets. The IPCC Third Assessment Report gave an estimated range for the radiative forcing associated with the first indirect aerosol effect (0 to -2 W m^{-2}); this range was larger than the uncertainty attributed to any of the other forcings, reflecting in large part the very low level of scientific understanding. Potential magnitudes of the second indirect effect and the semi-direct effect were not estimated in the report.

A number of research avenues hold promise for improving understanding of indirect and semidirect aerosol effects and better constraining estimates of their magnitudes. These include climate modeling, laboratory measurements, field campaigns, and satellite measurements. To improve the representation of the indirect effect in climate models, fundamental research is needed on the physical and chemical composition of aerosols, aerosol activation, cloud microphysics, cloud dynamics, and subgrid-scale variability in relative humidity and vertical velocity.

RECOMMENDATIONS:

•   Conduct integrated and comprehensive field investigations to quantify indirect aerosol radiative forcings—for example, in closure experiments with redundant observational and modeling studies.

•   Enhance the value of information derived from satellite instruments with targeted field campaigns and greater support for analysis of long-term surface records.

❖   Improve understanding and parameterizations of the indirect aerosol radiative and nonradiative effects in GCMs using process models, laboratory measurements, field campaigns, and satellite measurements.

•   Report the different indirect aerosol radiative forcings in climate change assessments and provide better estimates of the associated uncertainties.

Better Quantify the Direct Radiative Effects of Aerosols

Aerosols have direct radiative effects in that they scatter and absorb radiation. Knowledge of direct radiative forcing of aerosols is limited to a large extent by uncertainty in the global distribution and mixing states of the aerosols and by the role of different sources in contributing to atmospheric concentrations. Mixing states have major implications for aerosol optical properties that are not well understood and are difficult to parameterize in climate models. Another factor of uncertainty in representing aerosol direct radiative forcings in climate models is the small-scale variability of humidity and temperature, which has a major impact on aerosol optical properties. Describing the rapid growth of particles as humidities approach 100 percent is a particular challenge. Radiative transfer models relating aerosol columns and optical properties to the corresponding radiative forcing are thought to be relatively mature but must be tested further with field closure studies that provide multiple constraints for the models for a range of environments.

Assessments of past and future radiative forcings are compromised by the poor characterization of aerosol sources and sinks. Many natural and anthropogenic mechanisms of aerosol production are not understood: Their variability with future changes in population, technology, and climate cannot be accurately predicted. This finding is especially true for sea salt, dust, biomass burning, and the sources of carbonaceous aerosol. Removal of aerosols from the atmosphere occurs mainly by wet deposition, but model parameterizations of this process are highly uncertain and rudimentary in their coupling to the hydrological cycle.

RECOMMENDATIONS:

- Improve understanding of the global distribution of aerosols and their relationship to sources using data assimilation and inverse modeling approaches.

- Improve understanding of the radiative forcing of aerosols by direct measurements of radiative fluxes at the surface and TOA, vertical profiles of aerosols, their scattering and absorption coefficients, and their hygroscopic growth factors.

❖ Improve representation in global models of aerosol microphysics, growth, reactivity, and processes for their removal from the atmosphere through laboratory studies, field campaigns, and process models.

❖ Better characterize the sources and the physical, chemical, and optical properties of carbonaceous and dust aerosols.

- Improve estimates of black carbon and organic carbon emissions for the past (last 100 years) and future (next 100 years), utilizing indicators beyond static population density.

Better Quantify Radiative Forcing by Ozone

Ozone is a major greenhouse gas. The greatest uncertainty in quantifying this forcing lies in reconstructing the ozone concentration field in the past and projecting it into the future. Simulations of ozone with chemical transport models involve complex photochemical mechanisms coupled to transport on all scales and remain a major challenge, particularly in the troposphere. The inability of models to reproduce ozone trends over the twentieth century is disturbing and suggests that there could be large errors in current estimates of natural ozone levels and the sensitivity of ozone to human influence. In the troposphere these errors could relate to emissions of precursors, heterogeneous and homogeneous chemical processes, and stratospheric influence. Lightning emissions of nitrogen oxides (NO_x) are particularly uncertain and yet play a major role for ozone production in the middle and upper troposphere where the radiative effect is maximum. Transport of ozone between the stratosphere and troposphere greatly affects upper tropospheric concentrations in a manner that is still poorly understood.

RECOMMENDATIONS:
- Improve knowledge of sources of ozone precursors, including in particular nitrogen oxide emissions from lightning.
- Improve understanding of the transport of ozone in the upper troposphere and lower stratosphere region and the ability of models to describe this transport.
- Use observed long-term century trends in ozone concentrations to evaluate and improve global chemical transport models.

Investigate the Potential for Abrupt Climate Change

Paleoclimate records indicate that climate can change so rapidly and unexpectedly that human or natural systems may have difficulty adapting. Such abrupt climate changes take place when "the climate system is forced to cross some threshold, triggering a transition to a new state at a rate determined by the climate system itself and faster than the cause" (NRC, 2002). The Earth's climate has experienced abrupt shifts in temperature and precipitation during the preindustrial Holocene. Each of these events appears to have weakened the North Atlantic thermohaline circulation enough to cause abrupt cooling of the northern North Atlantic, Greenland, Iceland, and Europe.

The present climate could undergo abrupt changes in the future, not necessarily by the same mechanisms as in the past. Models imply, for example, that greenhouse warming may alter the hydrologic cycle enough to freshen North Atlantic surface waters and shift thermohaline circulation closer to a threshold. Collapse of parts of the Greenland ice sheet could be a risk factor as suggested by evidence that meltwater-induced basal sliding of southern parts of the ice sheet toward the ocean may have begun within the last decade. Knowledge of what triggers abrupt climate changes is still quite limited; more research is needed to determine the possible role of radiative and nonradiative climate forcings, such as human-caused increases in greenhouse gases or land-use changes. Indeed, past abrupt climate changes have been especially common when the climate system itself was being altered.

The current understanding of abrupt climate change is discussed in a recent National Research Council report *Abrupt Climate Change: Inevitable Surprises* (NRC, 2002). That report provides more detailed recommendations for research needed to improve our understanding of and ability to predict possible abrupt changes in the future, including enhanced research on possible causes of abrupt change. Here, a few recommendations that pertain specifically to radiative forcing are identified. The committee notes that the recommended research will require long-term efforts.

RECOMMENDATIONS:

  • Investigate the magnitude, spatial patterns, and temporal variation of radiative forcing that may cause the climate system to cross a threshold (e.g., shutdown of the thermohaline circulation).

  • Conduct societal impact studies to investigate the magnitude of future forcing that would cause a crossing of a threshold in societal vulnerability.

  • Determine the probability that future radiative and nonradiative forcings (e.g., reductions in aerosol emissions, continued tropical deforestation) could induce an abrupt climate change.

IMPROVING THE OBSERVATIONAL RECORD

The most important step for improving understanding of forcings is to obtain a robust record of radiative forcing variables, both in the past and into the future. A robust observational record is essential for improved understanding of the past and future evolution of climate forcings and responses. Existing observational evidence from surface-based networks, other in situ data (e.g., aircraft campaigns, ocean buoys), remote sensing platforms, and a range of proxy data sources (e.g., tree rings, ice cores) has enabled substantial progress in understanding, but important shortcomings remain. The observational evidence needs to be more complete both in terms of the spatial and spectral coverage and in terms of the quantities measured. Long-term monitoring of forcing and climate variables at much improved accuracy is necessary to detect and understand future changes.

Advance the Attribution of Decadal to Centennial Climate Change

Carefully attributing past climate changes to known natural and anthropogenic forcings provides information on how such forcings may impact large-scale climate in the future. Instrumental records of past climate conditions and of the magnitude of various forcings extend back about 150 years at best. Comparisons of observed surface temperatures with those simulated using reconstructions of the past forcings have yielded important insights into the roles of various natural and anthropogenic factors governing climate change. The shortness of the instrumental record and of accurate monitoring of climate forcings, however, limits the confidence with which climate change since preindustrial times can be attributed to specific forcings. Proxy records obtained from ice cores, sediments, tree rings, and other sources provide a critical tool for extending knowledge of both climate forcings and climate response further back in history. Improved historical radiative forcing reconstructions will require new understanding of physical process to better understand the relevance of available historical observations and their relationship to the actual forcings. For example, in the case of solar forcing, separate physical connections must be made of the solar magnetic activity with irradiance and with the near-Earth space environment, which modulates galactic cosmic rays that produce cosmogenic isotopes—the only long-term archive of solar activity and a crucial component of long-term Sun-climate research.

The lack of proxy climate data in certain key regions (e.g., large parts of the tropical Pacific and the extratropical oceans) is a major limitation. Such regional information is important in evaluating the potential roles of changes in modes of climate variability, such as ENSO, in past climate changes. Experiments employing fully coupled land-ocean-atmosphere models to study regional past climate change are just now under way. For comparison with model simulations, greater historical knowledge should be sought for a broad array of climate system parameters including the hydrological cycle (e.g., droughts, rainfall, streams), modes of variability (e.g., ENSO, annular modes), land use, the stratosphere, and ozone.

RECOMMENDATIONS:

*   Seek greater historical knowledge for a broad array of climate system parameters including the hydrological cycle (e.g., droughts, rainfall, streams), modes of variability (e.g., ENSO, annular modes), land use, the stratosphere, and ozone.
*   Improve proxy records of past radiative forcing (e.g., indicators of solar and volcanic forcing).
*   Enhance physical understanding of how these proxy records relate to the forcings (e.g., relationship between solar activity, irradiance, and cosmogenic isotopes).
*   Undertake "data archeology" projects to recover long-term instrumental records of climate variables of the past few centuries.
*   Continue to develop high-quality, high-resolution (or well-dated,

lower-resolution) proxy records of past climate change, and synthesize these data into spatially and temporally resolved reconstructions of climate change in past centuries to millennia.

❖ Develop a best-estimate climate forcing history for the past century to millennium.

❖ Using an ensemble of climate models, simulate the regional and global climate response to the best-estimate forcings and compare to the observed climate record.

Conduct Accurate Long-Term Monitoring of Radiative Forcing Variables

Geophysical quantities relevant to climate forcing should be known with a level of accuracy that is significantly smaller than the expected changes. The current approach relies primarily on measurement repeatability (precision), using overlapping successive measurements to cross-calibrate their absolute uncertainties. The principles described in Karl et al. (1995) (endorsed in *Adequacy of Climate Observing Systems*; NRC, 1999) provide a suitable framework for guiding collection of observations of radiative forcing and other climate variables. These principles have been updated in the *Strategic Plan for the U.S. Climate Change Science Program* (USCCSP, 2003).

Ultimately, the measurement accuracies of the geophysical parameters must be tied to irrefutable absolute standards and be tested and validated in perpetuity. Such benchmark measurements of radiative and other climate forcings and climate variables are needed immediately. Because the radiative forcings and the climate responses are highly dependent on wavelength, space-based observations with high spectral resolution are needed to isolate the signatures of the relevant radiative processes and components. Because the forcings and responses that determine any one particular climate state involve a distribution about a mean, the ensemble must be properly characterized and quantified so that changes in the mean can be reliably identified. Ultimately, the specification of the forcings and responses must be integrated to test climate forecast models.

Observational networks for the detection of long-term changes in climate variables must be improved. For example, local land-use changes and vegetation dynamics (i.e., microclimate effects) have been shown for some long-term climate monitoring sites to result in surface air temperature trends that are not spatially representative. Photographic and other documentation of monitoring sites and the surrounding landscape is needed to document the integrity of the sites over time.

Surface and tropospheric heat content changes may provide in the future a robust evaluation of climate changes. Long-term, globally averaged changes in the heat content of the oceans permit the calculation of the globally averaged radiative forcing over the timescale of the averaging period. Ocean heat storage changes have been shown to be an essential metric that climate model simulations must skillfully reproduce. The accurate assessment of concurrent heat storage changes in the atmosphere, land, and continental glaciers and sea ice would permit the averaging time to be shorter. Measurements of moist enthalpy can be used to characterize the heat content in surface air, providing more information about the surface energy budget than given by surface air temperature alone. A network of surface stations intended to characterize the surface energy budget could help better understand and monitor nonradiative forcings, although care would be needed in determining the siting and density of stations to appropriately account for the impact of landscape heterogeneity.

RECOMMENDATIONS:

❖ Continue observations of climate forcings and variables without interruption for the future in a manner consistent with established climate monitoring principles (e.g., adequate cross-calibration of successive, overlapping datasets).

❖ Develop the capability to obtain benchmark measurements (i.e., with uncertainty significantly smaller than the change to be detected) of key parameters (e.g., sea level altimetry, solar irradiance, and spectrally resolved, absolute radiance to space).

• Continuously monitor key radiative forcing parameters with high spectral resolution in order to isolate and understand the physical processes (e.g., solar spectral irradiance; surface, ocean, and atmosphere radiance to space), and ensure continuity and radiometric compatibility with existing and future broadband satellite measurements of shortwave and longwave irradiance.

• Conduct a comprehensive review and documentation of current and historical surface observation sites that are used in long-term temperature monitoring.

❖ Conduct highly accurate measurements of global ocean heat content and its change over time.

• Explore the value of creating a network of surface sites that provide representative monitoring of the surface energy budget.

ADDRESSING POLICY NEEDS

The concept of radiative forcing has clear policy applications. It has been used by the policy community to compare different forcings and as input to simple climate models used to consider policy options. Control strategies designed to address other environmental issues, such as air pollution or land-use changes, can also impact radiative forcings, a consequence that is rarely considered in developing such strategies. One specific concern is the possibility that reducing aerosol concentrations could enhance radiative warming. In addition, the policy community has focused primarily on global mean radiative forcing and the associated response in surface temperature. Given the increasing realization of the significance of geographically dependent climate forcings, the policy community will need new forcing metrics and guidance on how to apply them.

Integrate Climate Forcing Criteria in Environmental Policy Analysis

Policies designed to manage air pollution and land use may be associated with unintended impacts on climate. For example, aerosol and ozone have significant impacts on human health, ecosystems, and climate. Emissions of aerosols, aerosol precursors, and ozone precursors are already regulated in the United States and other industrialized nations. Increasing evidence of their health effects makes it likely that aerosols will be the target of further regulations to reduce their concentrations in the future. To date, these control strategies have not considered the potential climatic implications of emissions reductions. Regulations targeting black carbon emissions or ozone precursors would have combined benefits for public health and climate. However, because some aerosols have a negative radiative forcing, reducing their concentrations could actually accelerate radiative warming. Understanding of the effect of aerosols on the hydrological cycle and vegetation is still incomplete, making it is difficult to predict the total effect on climate of reducing aerosol emissions.

The assumptions made about the magnitude of climate sensitivity are an important consideration associated with regulations that attempt to reduce aerosols. Several modeling studies have suggested that aerosol direct and indirect forcing may have offset as much as 50 to 75 percent of the greenhouse gas forcing since preindustrial times. At the same time, the IPCC Third Assessment Report and climate modeling studies attribute the large warming witnessed during the recent decades to the increase of concentrations of carbon dioxide (CO_2) and other greenhouse gases. These two findings taken together reveal the possibility that climate sensitivity due to radiative forcing is in the upper range of the 1.5 to 4.5 K global-averaged surface warming for a doubling of CO_2. This implies that attempts to regulate air pollution, which would reduce aerosol abundances, could inadvertently trigger a strong acceleration of global surface warming in the coming decades.

Policies associated with land management practices also need to consider their inadvertent effects on climate. The continued conversion of landscapes by human activity, particularly in the humid tropics, could have complicated and possibly important consequences for regional and global climate change as a result of changes in the surface energy budget.

RECOMMENDATIONS:
- • Improve projections of future emissions of aerosols, aerosol precursors, and ozone precursors.
- • Improve projections of future land-use changes.
- ❖ Apply climate models to the investigation of scenarios in which aerosols are significantly reduced over the next 10 to 20 years and for a range of cloud microphysics parameterizations.
- ❖ Integrate climate forcing criteria in the development of future policies for air pollution control and land management.

Provide Improved Guidance to the Policy Community

The radiative forcing concept is used to inform climate policy discussions, in particular to compare the expected relative impacts of forcing agents. For example, integrated assessment models use radiative forcing as input to simple climate models, which are linked with socioeconomic models that predict the economic damages from climate impacts and the costs of various response strategies. This approach has been used to evaluate potential greenhouse gas emissions control strategies for meeting the Kyoto Protocol targets, as well as other policy questions. Many simplified climate models have focused on global mean surface temperature as the primary climate response to forcings, although more recently they have considered regional temperature changes and other societally relevant aspects of climate, such as sea level. It is important that models used for policy analysis incorporate further complexities in the radiative and nonradiative forcing concepts, as identified in this chapter. It is important also to communicate the expanded forcing concepts as described in this report to the policy community and to develop the tools that will make their application useful in a policy context.

Many climate policy questions require comparing the climate change effects of different greenhouse gases, aerosols, and other forcings. The concept of global warming potential (GWP) was developed to address this need. Application of the GWP concept has been restricted mainly to the long-lived greenhouse gases. In principle, it could be applied to short-lived forcing agents such as ozone and aerosols or, more specifically, to the emissions of their precursors, but there are a number of complicating factors including (1) the often poorly defined relationship between a precursor and a radiative forcing agent; (2) the inhomogeneity of the forcing; and (3) the much shorter time horizons (decades or less) relevant to the radiative forcing from these short-lived agents. In addition, the current concept is not useful for evaluating how the rate of technical transformation, which depends on economic and policy drivers, affects the trade-off between two greenhouse gases.

For most policy applications, the relationship between radiative forcing and temperature is assumed to be linear, suggesting that radiative forcing from individual positive and negative forcing agents could be summed to determine a net forcing. This assumption is generally reasonable for homogeneously distributed greenhouse gases, but it does not hold for all forcings. Thus, the assumed linearity of radiative forcing has been simultaneously useful and misleading for the policy community. It is important to determine the degree to which global mean TOA forcings are additive and whether one can expect, for example, canceling effects on climate change from changes in greenhouse gases on the one hand and changes in reflective aerosols on the other.

RECOMMENDATIONS:

❖   Encourage policy analysts and integrated assessment modelers to move beyond simple climate models based entirely on global mean TOA radiative forcing and incorporate new global and regional radiative and nonradiative forcing metrics as they become available.

•   Devise practical tools to relate new forcing metrics that may be introduced in the future to simple measures of climate change.

•   Explore ways to extend the GWP concept to account for aerosols and aerosol precursors, regional variation in forcing, and economic and policy factors that might affect the long-term impact of forcings.

•   Provide guidance to policy analysts on how individual radiative forcings combine to produce a net radiative forcing with an associated uncertainty.

References

Abdul-Razzak, H., and S. J. Ghan. 2000. A parameterization of aerosol activation: 2. Multiple aerosol types. Journal of Geophysical Research 105:6837-6844.

Adams, J. B., M. E. Mann, and C. M. Ammann. 2003. Proxy evidence for an El Niño-like response to volcanic forcing. Nature 426:274-278.

Albrecht, B. A. 1989. Aerosols, cloud microphysics, and fractional cloudiness. Science 245:1227-1230.

Alcamo, J., ed. 1994. IMAGE 2.0: Integrated Modeling of Global Climate Change. Dordrecht, The Netherlands: Kluwer Academic Publishers.

Allen, M. R., P. A. Stott, J. F. B. Mitchell, R. Schnur, and T. L. Delworth. 2000. Quantifying the uncertainty in forecasts of anthropogenic climate change. Nature 407:617-620.

Ambartsumyan, V. A. 1958. Theoretical Astrophysics. London: Pergamon Press.

Ammann, C. M., G. A. Meehl, W. M. Washington, and C. S. Zender. 2003. A monthly and latitudinally varying volcanic forcing dataset in simulations of the 20th century climate. Geophysical Research Letters 30:1657, DOI: 10.1029/2003GL016875

Ammann, C. M., G. A. Meehl, W. M. Washington, and C. S. Zender. Submitted. Climate simulations of the 20th century with the PCM. Geophysical Research Letters.

Anderson, T. L., R. J. Charlson, S. E. Schwartz, R. Knutti, O. Boucher, H. Rodhe, and J. Heintzenberg. 2003a. Climate forcing by aerosols—A hazy picture. Science 300:1103-1104, DOI: 10.1126/science.1084777.

Anderson, T. L., R. J. Charlson, D. M. Winker, J. A. Ogren, and K. Holmén. 2003b. Mesoscale variations of tropospheric aerosols. Journal of the Atmospheric Sciences 60(1):119-136.

Andronova, N. G., and M. E. Schlesinger. 2001. Objective estimation of the probability density function for climate sensitivity. Journal of Geophysical Research 106(D19):22605-22612, DOI: 10.1029/2000JD000259.

Annual Energy Outlook. 2004. Annual Energy Outlook 2004: With Projections to 2025. Energy Information Administration, U.S. Department of Energy, DOE/EIA-0383(2004), also available at *http://www.eia.doe.gov/oiaf/aeo/*.

Arellano, A. F., Jr., P. S. Kasibhatla, L. Giglio, G. R. van der Werf, and J. T. Randerson. 2004. Top-down estimates of global CO sources using MOPITT measurements. Geophysical Research Letters 31, L01104, DOI: 10.1029/2003GL018609.

Avissar, R., and D. Werth. 2005. Global hydroclimatological teleconnections resulting from tropical deforestation. Journal of Hydrometeorology, in press.

Bahreini, R., J. L. Jimenez, J. Wang, R. C. Flagan, J. H. Seinfeld, J. T. Jayne, and D. Worsnop. 2003. Aircraft-based aerosol size and composition measurements during ACE-Asia using an aerodyne aerosol mass spectrometer. Journal of Geophysical Research 108(D23): 8645, DOI: 10.1029/2002JD003226.

Baldocchi, D. D., E. Falge, L. Gu, R. Olson, D. Hollinger, S. Running, P. Anthoni, Ch. Bernhofer, K. Davis, J. Fuentes, A. Goldstein, G. Katul, B. Law, X. Lee, Y. Malhi, T. Meyers, J. W. Munger, W. Oechel, K. Pilegaard, H. P. Schmid, R. Valentini, S. Verma, T. Vesala, K. Wilson, and S. Wofsy. 2001. FLUXNET: A new tool to study the temporal and spatial variability of ecosystem-scale carbon dioxide, water vapor and energy flux densities. Bulletin of the American Meteorological Society 82:2415-2435.

Baldwin, M. P., and T. J. Dunkerton. 1999. Propagation of the Arctic Oscillation from the stratosphere to the troposphere. Journal of Geophysical Research 104:30937.

Baliunas, S. L., R. A. Donahue, W. H. Soon, J. H. Horne, J. Frazer, L. Woodard-Eklund, M. Bradford, L. M. Rao, O. C. Wilson, Q. Zhang, W. Bennett, J. Briggs, S. M. Carroll, D. K. Duncan, D. Figueroa, H. H. Lanning, T. Misch, J. Mueller, R. W. Noyes, D. Poppe, A. C. Porter, C. R. Robinson, J. Russell, J. C. Shelton, T. Soyumer, A. H. Vaughan, and J. H. Whitney. 1995. Chromospheric variations in main-sequence stars. Astrophysical Journal 438:269-287.

Bard, E., G. Raisbeck, P. Yiou, and J. Jouzel. 2000. Solar irradiance during the last 1200 years based on cosmogenic nuclides. Tellus B 52:985-992.

Barnett, T. P., D. W. Pierce, and R. Schnur. 2001. Detection of anthropogenic climate change in the world's oceans. Science 292:270-274.

Barrie, L. A., P. Borrell, O. Boucher, J. Burrows, C. Camy-Peyret, J. Fishman, A. Goede, C. Granier, E. Hilsenrath, D. Hinsman, H. Kelder, J. Langen, V. Mohnen, T. Ogawa, T. Peter, P. C. Simon, P.-Y. Whung, and A. Volz-Thomas. 2004. International Global Atmospheric Chemistry Observations System (IGACO) Theme Report, L. Barrie, P. Borrell, and J. Langen, eds. Newcastle-under-Lyme, UK: P & PMB Consultants. Available on-line at *http://www.luna.co.uk/~pborrell/p&pmb/IGACO/IGACO-report.htm.*

Bartalev, S. A., V. G. Bondur, A. A. Gitelson, C. Justice, E. A. Loupian, D. Cline, V. I. Gorny, T. E. Khromova, P. W. Stackhouse, and S. V. Victorov. 2004. Remote sensing of the Northern Eurasia. Chapter 4 in NEESPI Science Plan. Available online at *http:// neespi.gsfc.nasa.gov/science/science.html.* Accessed January 27, 2005.

Bauer, E., M. Claussen, and V. Brovkin. 2003. Assessing climate forcings of the Earth system for the past millennium. Geophysical Research Letters 30, DOI: 10.1029/ 2002GL016639.

Beer, J., A. Blinov, G. Bonani, R. C. Finkel, H. J. Hofmann, B. Lehmann, H. Oeschger, A. Sigg, J. Schwander, T. Staffelbach, B. Stauffer, M. Suter, and W. Wötfli. 1990. Use of ^{10}Be in polar ice to trace the 11-year cycle of solar activity. Nature 347:164-166, DOI: 10.1038/347164a0.

Beer, R., T. A. Glavich, and D. M. Rider. 2001. Tropospheric emission spectrometer for the Earth Observing System's Aura satellite. Applied Optics 40:2356-2367.

Bengtsson, L., K. I. Hodges, and S. Hagemann. 2004. Sensitivity of large-scale atmospheric analyses to humidity observations and its impact on the global water cycle and tropical and extratropical weather systems in ERA40. Tellus A 56(3):202.

Berner, R. A., and Z. Kothavala. 2001. GEOCARB III: A revised model of atmospheric CO_2 over Phanerozoic time. American Journal of Science 301:182-204.

Berner, R. A., A. C. Lasaga, and R. M. Garrels. 1983. The carbonate-silicate geochemical cycle and its effect on atmospheric carbon dioxide over the past 100 million years. American Journal of Science 283:641-683.

Bertrand C., M. F. Loutre, M. Crucifix, and A. Berger. 2002. Climate of the last millennium: A sensitivity study. Tellus 54(A):221-244.

Betts, A. K. 2004. Understanding hydrometeorology using global models. Bulletin of the American Meteorological Society 85(11):e, DOI: 10.1175/BAMS-85-11-1673.

Betts, R. A. 2001. Biogeophysical impacts of land use on present-day climate: near-surface and radiative forcing. Atmospheric Science Letters, DOI: 10.1006/asle.2001.0023.

Bianchi, G. G., and I. N. McCave. 1999. Holocene periodicity in North Atlantic climate and deep-ocean flow south of Iceland. Nature 397:515-517.

Blunier, T., J. Chappellaz, J. Schwander, B. Stauffer, and D. Raynaud. 1995. Variations in atmospheric methane concentration during the Holocene epoch. Nature 374:46-49.

Boer, G. J., G. Flato, M. C. Reader, and D. Ramsden. 2000. A transient climate change simulation with greenhouse gas and aerosol forcing: Experimental design and comparison with the instrumental record for the 20th century. Climate Dynamics 16:405-425.

Bond, G. C., B. Kromer, J. Beer, R. Muschler, M. N. Evand, W. Showers, S. Hoffmann, R. Lotti-Bond, I. Hajdas, and G. Bonani. 2001. Persistent solar influence on North Atlantic climate during the Holocene. Science 294:2130-2138.

Bond, T. C., D. G. Streets, K. F. Yarber, S. M. Nelson, J.-H. Woo, and Z. Klimont. 2004. A technology-based global inventory of black and organic carbon emissions from combustion. Journal of Geophysical Research 109:14203, DOI: 10.1029/2003JD003697.

Bongaarts, J., and R. A. Bulatao. 2000. Beyond Six Billion: Forecasting the World's Population. Washington, D.C.: National Research Council.

Boucher, O. 1999. Air traffic may increase cirrus cloudiness. Nature 397:30-31.

Boucher, O., and J. Haywood. 2001. On summing the components of radiative forcing of climate change. Climate Dynamics 18:297-302.

Boucher, O., and U. Lohmann. 1995. The sulfate-CCN-cloud albedo effect: A sensitivity study using two general circulation models. Tellus 47B:281-300.

Bowman, K. W., T. Steck, H. M. Worden, J. Worden, S. Clough, and C. D. Rodgers. 2002. Capturing time and vertical variability of tropospheric ozone: A study using TES nadir retrieval. Journal of Geophysical Research 107(D23):4723, DOI: 10.1029/2002JD002150.

Breon, F.-M., D. Tanre, and S. Generoso. 2002. Aerosol effect on cloud droplet size monitored from satellite. Science 295:834-838.

Briffa, K. R., P. D. Jones, F. H. Schweingruber, and T. J. Osborn. 1998. Influence of volcanic eruptions on Northern Hemisphere summer temperatures over 600 years. Nature 393:450-455.

Broecker, W. S. 1994. Terminations. Pp. 687-698 in Milankovitch and Climate: Understanding the Response to Astronomical Forcing, A. L. Berger, J. Imbrie, J. Hays, G. Kukla, and B. Saltzman, eds. Dordrecht, The Netherlands: D. Rieidel.

Broecker, W. S. 1997. Thermohaline circulation, the Achilles heel of our climate system: Will man-made CO_2 upset the current balance? Science 278(5343):1582-1588.

Brühl, C. 1993. The impact of the future scenarios for methane and other chemically active gases on the GWP of methane. Chemosphere 26:731-738.

Budyko, M. I. 1969. The effect of solar radiation variations on the climate of the Earth. Tellus 21:611.

Cane, M. A., A. C. Clement, A. Kaplan, Y. Kushnir, D. Pozdnyakov, R. Seager, S. E. Zebiak, and R. Murtugudde. 1997. Twentieth-century sea surface temperature trends. Science 275:957-960.

Carslaw, K. S., R. G. Harrison, and J. Kirkby. 2002. Cosmic rays, clouds, and climate. Science 29:1732-1737.

Carton, J. A., G. Chepurin, X. Cao, and B. Giese. 2000. A simple ocean data assimilation analysis of the global upper ocean 1950-95. Part I. Methodology. Journal of Physical Oceanography 30:294-309.

Catling, D. C., K. J. Zahnle, and C. P. McKay. 2001. Biogenic methane, hydrogen escape, and the irreversible oxidation of early Earth. Science 293:839-843.

CDIAC (Carbon Dioxide Information Analysis Center). 2004. Oak Ridge National Laboratories, Oak Ridge, Tenn., Available at: *http://cdiac.esd.ornl.gov/*.

Cess, R. D. 1972. The thermal structure within the stratospheres of Venus and Mars. Icarus 17(2):561-569.

Cess, R. D. 1976. Climate change: An appraisal of atmospheric feedback mechanisms employing zonal climatology. Journal of the Atmospheric Sciences. 33(10):1831-1843.

Cess, R. D., and G. L. Potter. 1988. A methodology for understanding and intercomparing atmospheric climate feedback processes in GCMs. Journal of Geophysical Research 93: 8305-8314.

Chamberlain, J. W. 1960. Interplanetary gas. II: Expansion of a model solar corona. Astrophysics Journal 131(47).

Chameides W. L., H. Hu, S. C. Liu, M. Bergin, X. Zhou, L. Mearns, G. Wang, C. S. Kiang, R. D. Saylor, C. Luo, Y. Huang, A. Steiner, and F. Giorgi. 1999. Case study of the effects of atmospheric aerosols and regional haze on agriculture: An opportunity to enhance crop yields in China through emission controls? Proceedings of the National Academy of Sciences 26:13626-13633.

Chandrasekhar, S. 1947. The transfer of radiation in stellar atmospheres. Josiah Willard Gibbs Lecture. Bulletin of the American Mathematical Society 53:641.

Charlock, T. P., and V. Ramanathan. 1985. The albedo field and cloud radiative forcing produced by a general circulation model with internally generated cloud optics. Journal of the Atmospheric Sciences 42:1405-1429.

Charlson, R. J., J. Langner, H. Rodhe, C. B. Leovy, and S. G. Warren. 1991. Perturbation of the Northern Hemisphere radiative balance by backscattering from anthropogenic sulfate aerosols. Tellus 43AB(4):152-163.

Chase, T. N., R. A. Pielke Sr., T. G. F. Kittel, R. R. Nemani, and S. W. Running. 2000a. Simulated impacts of historical land cover changes on global climate in northern winter. Climate Dynamics 16:93-105.

Chase, T. N., R. A. Pielke Sr., J. A. Knaff, T. G. F. Kittel, and J. L. Eastman. 2000b. A comparison of regional trends in 1979-1997 depth-averaged tropospheric temperatures. International Journal of Climatology 20:503-518.

Chase, T. N., R. A. Pielke, Sr., B. Herman, and X. Zeng. 2004. Likelihood of rapidly increasing surface temperatures unaccompanied by strong warming in the free troposphere. Climate Research 25:185-190.

Chen, C-T., and V. Ramaswamy. 1996. Sensitivity of simulated global climate to perturbations in low cloud microphysical properties. Part II: Spatially localized perturbations. Journal of Climate 9:2788-2801.

Chertow, M. 2001. The IPAT equation and its variants: Changing views of technology and environmental impact. Journal of Industrial Ecology 4(4).

Christy J. R., and W. B. Norris. 2004. What may we conclude about tropospheric temperature trends? Geophysical Research Letters 31(6):L0621.

Chung, C. E., and V. Ramanathan. 2003. South Asian haze forcing: Remote impacts with implications to ENSO and AO. Journal of Climate 16:1791-1806.

Chung, C. E., V. Ramanathan, and J. T. Kiehl. 2002. Effects of the south Asian absorbing haze on the northeast monsoon and surface-air heat exchange. Journal of Climate 15(17):2462-2476.

Chung, S. H., and J. H. Seinfeld. 2002. Global distribution and climate forcing of carbonaceous aerosols. Journal of Geophysical Research 107(D19):4407, DOI: 10.1029/2001JD001397.

Chylek, P., V. Ramaswamy, A. Ashkin, and J. M. Dziedzic. 1983. Simultaneous determination of refractive index and size of spherical dielectric particles from light scattering data. Applied Optics 22:2302-2307.

Chylek, P., G. Lesins, G. Videen, J. Wong, R. Pinnick, D. Ngo, and J. Klett. 1996. Black carbon and absorption of solar radiation by clouds. Journal of Geophysical Research 101:23365-23371.

Claussen, M., A. Ganopolski, H.-J. Schellnhuber, and W. Crameret. 2002. Earth system models of intermediate complexity: Closing the gap in the spectrum of climate models. Climate Dynamics 18:579-586.

Claussen, M., P. M. Cox, X. Zeng, P. Viterbo, A. C. M. Beljaars, R. A. Betts, H.-J. Boole, T. Chase, and R. Koster. 2004. The global climate. Chapter A.4 in Vegetation, Water, Humans and the Climate: A New Perspective on an Interactive System, P. Kabat, M. Claussen, P. A. Dirmeyer, J. H. C. Gash, L. Bravo de Guenni, M. Meybeck, R. Pielke Sr., C. J. Vörösmarty, R. W. A. Hutjes, and S. Lütkemeier, eds. Berlin: Springer Verlag.

Clement, A. C., R. Seager, and M. A. Cane. 2000. Suppression of El Niño during the mid-Holocene by changes in the Earth's orbit. Paleoceanography 15:731-737.

Clilverd, M. A., E. Clarke, H. Rishbeth, T. G. Clark, and T. Ulich. 2003. Solar activity level in 2100. Astronomy and Geophysics 44(5):5.20-5.22.

Cline, W. R. 1992. The Economics of Global Warming. Washington, D.C.: Institute for International Economics.

Clough, S. A., C. P. Rinsland, and P. D. Brown. 1995. Retrieval of tropospheric ozone from simulations of nadir spectral radiances as observed from space. Journal of Geophysical Research 100:16579-16593.

Coakley, J. A., and C. D. Walsh. 2002. Limits to the aerosol indirect radiative effect derived from observations of ship tracks. Journal of the Atmospheric Sciences 59(3):668-680.

Coakley, J. A., R. L. Bernstein, and P. A. Durkee. 1987. Effect of ship-stack effluents on cloud reflectivity. Science 237:1020-1022.

Cobb, K. M., C. D. Charles, H. Cheng, and R. L Edwards. 2003. El Niño-Southern Oscillation and tropical Pacific climate during the last millennium. Nature 424:271-276.

Cochrane M. A., A. Alencar, M. D. Schulze, C. M. Souza, Jr., D. C. Nepstad, P. Lefebvre, and E. A. Davidson. 1999. Positive feedbacks in the fire dynamic of closed canopy tropical forests. Science 284(5421):1832-1835.

Collins, M. 2000. The El Niño-Southern Oscillation in the Second Hadley Centre Coupled Model and its response to greenhouse warming. Journal of Climate 13:1299-1312.

Collins, W. D., A. Bucholtz, D. Lubin, P. Flatau, F. P. J. Valero, C. P. Weaver, and P. Pilewskie. 2000. Determination of surface heating by convective cloud systems in the central equatorial Pacific from surface and satellite measurements. Journal of Geophysical Research 105:14807-14821.

Collins W. D., P. J. Rasch, B. E. Eaton, D. W. Fillmore, J T Kiehl, C. T. Beck, and C. S. Zender. 2002. Simulation of aerosol distributions and radiative forcing for INDOEX: Regional climate impacts. Journal of Geophysical Research 107(D19):8028, DOI: 10.1029/2000JD000032.

Conant, W. C., J. H. Seinfeld, J. Wang, G. R. Carmichael, Y. Tang, I. Uno, P. J. Flatau, K. M. Markowicz, and P. K. Quinn. 2003. A model for the radiative forcing during ACE-Asia derived from CIRPAS Twin Otter and R/V Ronald H. Brown data and comparison with observations. Journal of Geophysical Research 108(D23):8661, DOI: 10.1029/ 2002JD003260.

Cooke, W. F., V. Ramaswamy, and P. Kasibhatla. 2002. A general circulation model study of the global carbonaceous aerosol distribution. Journal of Geophysical Research 107(D16), DOI: 10.1029/2001JD001274.

Cox, P. M., R. A. Betts, C. D. Jones, S. A. Spall, and I. J. Totterdell. 2000. Acceleration of global warming due to carbon-cycle feedbacks in a coupled climate model. Nature 408:184-187.

Crisp, D. R., M. Atlas, F.-M. Breon, L. R. Brown, J. P. Burrows, P. Ciais, B. J. Connor, S. C. Doney, I. Y. Fung, D. J. Jacob, C. E. Miller, D. O'Brien, S. Pawson, J. T. Randerson, P. Rayner, R. J. Salawitch, S. P. Sander, B. Sen, G. L. Stephens, P. P. Tans, G. C. Toon, P. O. Wennberg, S. C. Wofsy, Y. L. Yung, Z. Kuang, B. Chudasama, G. Sprague, B. Weiss, R. Pollock, D. Kenyon, and S. Schroll. 2004. The Orbiting Carbon Observatory (OCO) Mission. Advances in Space Research 34:700-709.

Crowley, T. J. 1998. Significance of tectonic boundary conditions for paleoclimate simulations. Pp. 3-17 in Tectonic Boundary Conditions for Climate Reconstructions, T. J. Crowley and K. Burke, eds. New York: Oxford University Press.

Crowley, T. J. 2000. Causes of climate change over the past 1000 years. Science 289:270-277.

Crowley, T. J., and K.-Y. Kim. 1999. Modeling the temperature response to forced climate change over the last six centuries. Geophysical Research Letters 26:1901-1904.

Crowley, T. J., S. K. Baum, K.-Y. Kim, G. C. Hegerl, and W. T. Hyde. 2003. Modeling ocean heat content changes during the last millennium. Geophysical Research Letters 30:1932 DOI: 10.1029/2003GL017801.

Cubasch, U., G. A. Meehl, G. J. Boer, R. J. Stouffer, M. Dix, A. Noda, C. A. Senior, S. C. B. Raper, and K. S. Yap. 2001. Projections of Climate Change. Pp. 525-582 in Climate Change 2001: The Scientific Basis, J. T. Houghton, Y. Ding, D. J. Griggs, M. Noguer, P. J. van der Linden, X. Dai, K. Maskell and C.A. Johnson, eds. Cambridge, U.K.: Cambridge University Press.

Decesari, S., M. C. Facchini, E. Matta, F. Lettini, M. Mircea, S. Fuzzi, E. Tagliavini, and J. P. Putaud. 2001. Chemical features and seasonal trend of water soluble organic compounds in the Po Valley fine aerosol. Atmospheric Environment 35:3691-3699.

deMenocal, P., J. Ortiz, T. Guilderson, and M. Sarnthein. 2000. Coherent high and low latitude climate variability during the Holocene warm period. Science 288:2198-2202.

Denning, A. S., M. Nicholls, L. Prihodko, I. Baker, P.-L. Vidale, K. Davis, and P. Bakwin. 2003. Simulated variations in atmospheric CO_2 over a Wisconsin forest using a coupled ecosystem-atmosphere model. Global Change Biology 9(9):1241-1250.

Dentener, F., M. van Weele, M. Krol, S. Houweling, and P. van Velthoven. 2003. Trends and inter-annual variability of methane emissions derived from 1979-1993 global CTM simulations. Atmospheric Chemistry and Physics 3:73-88.

de Vries, B., J. Bollen, L. Bouwman, M. den Elzen, M. Janssen, and E. Kreileman. 2000. Greenhouse gas emissions in an equity-, environment, and service-oriented world: An IMAGE-based scenario for the 21st century. Technological Forecasting and Social Change 63:2-3.

de Vries, H. J. M., J. G. J. Olivier, R. A. van den Wijngaart, G. J. J. Kreileman, and A. M. C. Toet. 1994. Model for calculating regional energy use, industrial production and greenhouse gas emissions for evaluating global climate scenarios. Water, Air, and Soil Pollution 76:79-131.

Dlugokencky, E. J., S. Houweling, L. Bruhwiler, K. A. Masarie, P. M. Lang, J. B. Miller, and P. P. Tans. 2003. Atmospheric methane levels off: Temporary pause or a new steady-state? Geophysical Research Letters 30(19), DOI: 10.1029/2003GL018126.

Doran, J. C., and S. Zhong. 2002. Comments on "Atmospheric disturbances caused by human modification of landscape." Bulletin of the American Meteorological Society 83:277-279.

Douglass, D. H., and B. D. Clader. 2002. Climate sensitivity of the Earth to solar irradiance. Geophysical Research Letters 29(16), DOI: 10.1029/2002GL015345.

Dutton, E. G., J. J. Michalasky, T. Stoffel, B. W. Forgan, J. Hickey, D. W. Nelson, T. L. Alberta, and I. Reda. 2001. Measurement of broadband diffuse solar irradiance using current commercial instrumentation with a correction for thermal offset errors. Journal of Atmospheric and Oceanic Technology 18:297-314.

Eastman, J. L., M. B. Coughenour, and R. A. Pielke Sr. 2001a. Does grazing affect regional climate? Journal of Hydrometeorology 2:243-253.

Eastman, J. L., M. B. Coughenour, and R. A. Pielke Sr. 2001b. The effects of CO_2 and landscape change using a coupled plant and meteorological model. Global Change Biology 7:797-815.

Edmonds, J., M. Wise, H. Pitcher, R. Richels, T. Wigley, and C. MacCracken. 1996a. An integrated assessment of climate change and the accelerated introduction of advanced energy technologies: An application of MiniCAM 1.0. Mitigation and Adaptation Strategies for Global Change 1(4):311-339.

Edmonds, J., M. Wise, R. Sands, R. Brown, and H. Kheshgi. 1996b. Agriculture, land-use, and commercial biomass energy. A preliminary integrated analysis of the potential role of biomass energy for reducing future greenhouse related emissions. PNNL-11155. Washington, D.C.: Pacific Northwest National Laboratories.

Edwards, D. P., C. M. Halvorson, and J. C. Gille. 1999. Radiative transfer modeling for the EOS Terra satellite Measurement of Pollution in the Troposphere (MOPITT) instrument. Journal of Geophysical Research 104(D14):16755-16776.

Ehrlich, P. R., and J. P. Holdren. 1971. Impact of population growth. Science 171:1212-1217.

Ellis, J. S., T. H. Vonder Haar, S. Levitus, and A. H. Oort. 1978. The annual variation in the global heat balance of the earth. Journal of Geophysical Research 83:1958-1962.

Elvidge, C. D., K. E. Baugh, K. E. Hobson, E. A. Kihn, H. W. Krochl, H. W. Davis, E. R. Davis, and D. Cocero. 1997. Satellite inventory of human settlements using nocturnal radiation emissions: A contribution for the global toolchest. Global Change Biology 3:387-395.

Evans, W. F. J., and E. Puckrin. 1999. A comparison of GCM models with experimental measurements of surface radiative forcing by greenhouse gases. Pp. 378-381 in Tenth Symposium on Global Change Studies. Dallas, Texas: American Meteorological Society.

Facchini, M. C., S. Decesari, M. Mircea, S. Fuzzi, and G. Loglio. 2000. Surface tension of atmospheric wet aerosol and cloud/fog droplets in relation to their organic carbon content and chemical composition. Atmospheric Environment 34:4853-4857.

Feichter, J., E. Roeckner, U. Lohmann, and B. Liepert. 2004. Nonlinear aspects of the climate response to greenhouse gas and aerosol forcing. Journal of Climate 17(12):2384-2398, DOI: 10.1175/1520-0442(2004)017

Feingold, G., and P. Chuang. 2002. Analysis of the influence of film-forming compounds on droplet growth: Implications for cloud microphysical processes and climate. Journal of the Atmospheric Sciences 59:2006-2018.

Feingold G., W. L. Eberhard, D. E. Veron, and M. Previdi. 2003. First measurements of the Twomey indirect effect using ground-based remote sensors. Geophysical Research Letters 30(6):1287, DOI: 10.1029/2002GL016633.

Fiore, A. M., D. J. Jacob, I. Bey, R. M. Yantosca, B. D. Field, A. C. Fusco, and J. G. Wilkinson. 2002. Background ozone over the United States in summer: Origin, trend, and contribution to pollution episodes. Journal of Geophysical Research 107(D15), DOI: 10.1029/2001JD000982.

Fishman, J., C. E. Watson, J. C. Larsen, and J. A. Logan. 1990. Distribution of tropospheric ozone determined from satellite data. Journal of Geophysical Research 95:3599-3617.

Folland, C. K., D. M. H. Sexton, D. J. Karoly, C. E. Johnson, D. P. Rowell, and D. E. Parker. 1998. Influences of anthropogenic and oceanic forcing on recent climate change. Geophysical Research Letters 25(3):353-356.

Folland, C. K., N. A. Rayner, S. J. Brown, T. M. Smith, S. S. P. Shen, D. E. Parker, I. Macadam, P. D. Jones, R. Jones, N. Nicholls, and D. M. H. Sexton. 2001. Global temperature change and its uncertainties since 1861. Geophysical Research Letters 28:2621-2624.

Fontenla, J. M., J. Harder, G. Rottman, T. Woods, G. M. Lawrence, and S. Davis. 2003. The signature of solar activity in the infrared spectral irradiance. Astrophysical Journal Letters 605(1 part 2):L85.

Forest, C. E., P. H. Stone, A. P. Sokolov, M. R. Allen, and M. D. Webster. 2002. Quantifying uncertainties in climate system properties with the use of recent climate observations. Science 295:113-117, DOI: 10.1126/science.1064419.

Forster, P. M. de F., and K. P. Shine. 1999. Stratospheric water vapour changes as a possible contributor to observed stratospheric cooling. Geophysical Research Letters 26:3309-3312.

Forster, P. M. de F., and K. Tourpali. 2001. The effect of tropopause height changes on the calculation of ozone trends and their radiative forcing. Journal of Geophysical Research 106:12241-12252.

Forster, P. M. de F., R. S. Freckleton, and K. P. Shine. 1997. On aspects of the concept of radiative forcing. Climate Dymanics 13:547-560.

Foukal, P., and L. Milano. 2001. A measurement of the quiet network contribution to solar irradiance variation. Geophysical Research Letters 28:883-886.

Fowler, L. D., D. A. Randall, and S. A. Rutledge. 1996. Liquid and ice cloud microphysics in the CSU general circulation model. Part I: Model description and simulated microphysical processes. Journal of Climate 9:489-529.

Freckleton, R. S., E. J. Highwood, K. P. Shine, O. Wild, K. S. Law, and M. G. Sanderson. 1998. Greenhouse gas radiative forcing: Effects of averaging and inhomogeneities in trace gas distribution. Quarterly Journal of the Royal Meteorological Society 124:2099-2127.

Free, M., and A. Robock. 1999. Global warming in the context of the Little Ice Age. Journal of Geophysical Research 104:19057-19070.

Friddell, J. R., T. Thunell, T. Guilderson, and M. Kashgarian. 2003. Increased north-east Pacific climate variability during the warm middle Holocene. Geophysical Research Letters 30(11):1560, DOI: 10.1029/2003GL017048.

Fridlind, A. M., and M. Z. Jacobson. 2003. Point and column aerosol radiative closure during ACE 1: Effects of particle shape and size. Journal of Geophysical Research 108(D3), DOI: 10.1029/2001JD001553.

Friedlingstein P., L. Bopp, P. Ciais, J.-L Dufresne, L. Fairhead, H. LeTreut, P. Monfray, and J. Orr. 2001. Positive feedback between future climate change and the carbon cycle. Geophysical Research Letters 28:1543-1546.

Fröhlich, C., and J. Lean. 2002. Solar irradiance variability and climate. Astronomische Nachrichten 323:203-212.

Frouin, R., and S. F. Iacobellis. 2002. Influence of phytoplankton on the global radiation budget. Journal of Geographic Research 107(D19):5-1-5-10.

Fu, C., T. Yasunari, and S. Lutkemeier. 2004. The Asian monsoon climate. Chapter A.8 in
    Vegetation, Water, Humans and the Climate: A New Perspective on an Interactive Sys-
    tem, P. Kabat, M. Claussen, P. A. Dirmeyer, J. H. C. Gash, L. Bravo de Guenni, M.
    Meybeck, R. Pielke, Sr., C. J. Vörösmarty, R. W. A. Hutjes, and S. Lütkemeier, eds.
    Berlin: Springer Verlag.
Fuglestvedt, J. S., I. S. A. Isaksen, and W.-C. Wang. 1996. Estimates of indirect global warm-
    ing potentials for CH_4, CO and NO_x. Climatic Change 34:405-437.
Fuller, K., W. Malm, and S. Kreidenweis. 1999. Effects of mixing on extinction by carbon-
    aceous particles. Journal of Geophysical Research 104:15941-15954.
Fusco, A. C., and J. A. Logan. 2003. Analysis of 1970-1995 trends in tropospheric ozone at
    Northern Hemisphere midlatitudes with the GEOS-CHEM model. Journal of Geophysi-
    cal Research—Atmospheres 108(D15):4449.
Gaffin, S. R. 1998. World population projections for greenhouse gas emissions scenarios.
    Mitigation and Adaptation Strategies for Global Change 3(2-4):133-170.
Gaffin, S. R. 2002. Fuzzy math on greenhouse gas. New York Times, February 19.
Galloway, J. N., F. J. Dentener, D. G. Capone, E. W. Boyer, R. W. Howarth, S. P. Seitzinger,
    C. P. Asner, C. C. Cleveland, P. A. Green, E. A. Holland, D. M. Karl, A. F. Michaels, J.
    H. Porter, A. R. Townsend, and C. J. Vorosmarty. 2004. Nitrogen cycles: Past, present
    and future. Biogeochemistry 70:153-226.
Ganopolski, A., C. Kubatzki, M. Claussen, V. Brovkin, and V. Petoukhov. 1998. The influ-
    ence of vegetation-atmosphere-ocean interaction on climate during the mid-Holocene.
    Science 280:1916-1919.
Garrett, T. J., L. M. Russell, V. Ramaswamy, S. F. Maria, and B. J. Huebert. 2003. Micro-
    physical and radiative evolution of aerosol plumes over the tropical North Atlantic
    Ocean. Journal of Geophysical Research 108(D1):4022, DOI: 10.1029/2002JD002228.
Gates, W. L., J. Boyle, C. Covey, C. Dease, C. Doutriaux, R. Drach, M. Fiorino, P. Gleckler,
    J. Hnilo, S. Marlais, T. Phillips, G. Potter, B. Santer, K. Sperber, K. Taylor, and D.
    Williams. 1998. An overview of the results of the Atmospheric Model Intercomparison
    Project (AMIP I). Bulletin of the American Meterological Society 73:1962-1970.
Geller, M. A., and S. P. Smyshlyaev. 2002. A model study of total ozone evolution 1979-
    2000—The role of individual natural and anthropogenic effects. Geophysical Research
    Letters 29(22):2048 DOI: 10.1029/2002GL015689.
Gerber, S., F. Joos, P. P. Brügger, T. F. Stocker, M. E. Mann, S. Sitch and M. Scholze. 2003.
    Constraining temperature variations over the last millennium by comparing simulated
    and observed atmospheric CO_2. Climate Dynamics 20:281-299.
Ghan, S. J., L. R. Leung, R. C. Easter, and H. Abdul-Razzak. 1997. Prediction of cloud
    droplet number in a general circulation model. Journal of Geophysical Research
    102(21):777-794.
Gierasch, P., and R. M. Goody. 1968. A study of the thermal and dynamical structure of the
    Martian lower atmosphere. Planetary and Space Science 16:615-636.
Goldstein, G. A., and L. A. Greening. 1999. Energy Planning and the Development of Carbon
    Mitigation Strategies: Using the MARKAL Family of Models. Washington, D.C.: Inter-
    national Resources Group (IRG).
Gonzalez-Rouco, F., H. von Storch, and E. Zorita. 2003. Deep soil temperature as proxy for
    surface air-temperature in a coupled model simulation of the last thousand years. Geo-
    physical Research Letters 30:2116, DOI: 10.1029/2003GL018,264.
Govindasamy, B., P. B. Duffy, and K. Caldeira. 2001. Land use changes and Northern Hemi-
    sphere cooling. Geophysical Research Letters 28:291-294.
Gregory, J. M., W. J. Ingram, M. A. Palmer, G. S. Jones, P. A. Stott, R. B. Thorpe, J. A. Lowe,
    T. C. Johns, and K. D. Williams. 2004. A new method for diagnosing radiative forcing
    and climate sensitivity. Geophysical Research Letters 31:L03205, DOI: 10.1029/2003
    GL018747.

Gregory, J. M., R. J. Stouffer, S. C. B. Raper, P. A. Stott, and N. A. Rayner. 2002. An observationally based estimate of the climate sensitivity. Journal of Climate 15(22):3117-3121.

Grove, J. M. 1988. The Little Ice Age. London: Methuen.

Grubler, A. 1998. A review of global and regional sulfur emissions scenarios. Mitigation and Adaptation Strategies for Global Change 3(2):383-418.

Grubler, A., N. Nakićenović, and D. G. Victor. 1999. Dynamics of Energy Technologies and Global Change, Energy Policy 27(5):247-280.

Gu, L., D. D. Baldocchi, S. C. Wofsy, J. W. Munger, J. J. Michalsky, S. P. Urbanski, and T. A. Boden. 2003. Science 299:2035-2038.

Guenther, A., C. Geron, T. Pierce, B. Lamb, P. Harley, and R. Fall. 2000. Natural emissions of non-methane volatile organic compounds, carbon monoxide, and oxides of nitrogen from North America. Atmospheric Environment 34:2205-2230.

Gultepe, I., G. A. Issac, W. R. Leaitch, and C. M. Banic. 1996. Parameterizations of marine stratus microphysics based on in situ observations: Implications for GCMS. Journal of Climate 9(2):345-357, DOI: 10.1175/1520-0442(1996)009<0345:POMSMB>2.0.CO;2.

Gurney, K. R., R. M. Law, A. S. Denning, P. J. Rayner, D. Baker, P. Bousquet, L. Bruhwiler, Y.-H. Chen, P. Ciais, S. Fan, I. Y. Fung, M. Gloor, M. Heimann, K. Higuchi, J. John, T. Maki, S. Maksyutov, K. Masarie, P. Peylin, M. Prather, B. C. Pak, J. Randerson, J. Sarmiento, S. Taguchi, T. Takahashi, and C.-W. Yuen. 2002. Towards robust regional estimates of CO_2 sources and sinks using atmospheric transport models. Nature 415:626-630.

Haggerty, B. M. 1996. Episodes of flood-basalt volcanism defined by $^{40}Ar/^{39}Ar$ age distributions: Correlations with mass extinctions? Journal of Undergraduate Science 3:155-164.

Haigh, J. D. 2003. The effects of solar variability on the Earth's climate. Philosophical Transactions Royal Society of London A 361:95-111.

Hansen, J., and L. Nazarenko. 2004. Soot climate forcing via snow and ice albedos. Proceedings of the National Academy of Sciences 101:423-428, DOI: 10.1073/pnas.2237157100.

Hansen, J., A. Lacis, R. Ruedy, and M. Sato. 1992. Potential climate impact of Mount Pinatubo eruption. Geophysical Research Letters 19:215-218.

Hansen, J., M. Sato, and R. Ruedy 1997. Radiative forcing and climate response. Journal of Geophysical Research 102:6831-6864.

Hansen J., M. Sato, R. Ruedy, A. Lacis, and V. Oinas. 2000. Global warming in the twenty-first century: An alternative scenario. Proceedings of the National Academy of Sciences of the United States of America 97(18):9875-9880.

Hansen, J., M. Sato, L. Nazarenko, R. Ruedy, A. Lacis, D. Koch, I. Tegen, T. Hall, D. Shindell, B. Santer, P. Stone, T. Novakov, L. Thomason, R. Wang, Y. Wang, D. Jacob, S. Hollandsworth, L. Bishop, J. Logan, A. Thompson, R. Stolarski, J. Lean, R. Willson, S. Levitus, J. Antonov, N. Rayner, D. Parker, and J. Christy. 2002. Climate forcings in Goddard Institute for Space Studies SI2000 simulations. Journal of Geophysical Research 107(D18):4347, DOI: 10.1029/2001JD001143.

Harshvardhan, S. E. Schwartz, C. M. Benkovitz, and G. Guo. 2002. Aerosol influence on cloud microphysics examined by satellite measurements and chemical transport modeling. Journal of the Atmospheric Sciences 59(3):714-725.

Haug, G. H., K. A. Hughen, D. M. Sigman, L. C. Peterson, and U. Rohl. 2001. Southward penetration of the intertropical convergence zone through the Holocene. Science 293:1304-1308.

Hauglustaine, D., C. Granicr, G. Brasseur, and G. Megie. 1994. The importance of atmospheric chemistry in the calculation of radiative forcing on the climate system. Journal of Geophysical Research 99:1173-1186.

Haywood, J., and V. Ramaswamy. 1998. Global sensitivity studies of the direct radiative forcing due to anthropogenic sulfate and black carbon aerosols. Journal of Geophysical Research 103:6043-6058.

Haywood, J., V. Ramaswamy, and B. Soden. 1999. Tropospheric aerosol climate forcing in clear-sky satellite observations over the oceans. Science 283:1299-1303.

Heald, C. L., D. J. Jacob, D. B. A. Jones, P. I. Palmer, J. A. Logan, D. G. Streets, G. W. Sachse, J. C. Gille, R. N. Hoffman, and T. Nehrkorn. 2004. Comparative inverse analysis of satellite (MOPITT) and aircraft (TRACE-P) observations to estimate Asian sources of carbon monoxide. Journal of Geophysical Research 109:D23306, DOI: 10.1029/2004JD005185.

Hegerl, G. C., K. Hasselmann, U. Cubasch, J. F. B. Mitchell, E. Roeckner, R. Voss, and J. Waszkewitz. 1997. Multi-fingerprint detection and attribution analysis of greenhouse gas, greenhouse gas-plus-aerosol and solar forced climate change. Climate Dynamics 13:613-634.

Hegerl, G. C., P. A. Scott, M. R. Allen, J. F. B. Mitchell, S. F. B. Tett, and U. Cubasch. 2000. Optimal detection and attribution of climate change: Sensitivity of results to climate model differences. Climate Dynamics 16:737-754.

Hegerl, G. C., T. J. Crowley, S. K. Baum, K-Y. Kim, and W. T. Hyde. 2003. Detection of volcanic, solar and greenhouse gas signals in paleo-reconstructions of Northern Hemispheric temperature. Geophysical Research Letters 30:1242, DOI: 10.1029/2002GL016635.

Hein, R., P. Crutzen, and M. Heimann. 1997. An inverse modeling approach to investigate the global atmospheric methane cycle. Global Biogeochemical Cycles 11:43-76.

Hessler, A. M., D. R. Lowe, R. L. Jones, and D. K. Bird. 2004. A lower limit for atmospheric carbon dioxide levels 3.2 billion years ago. Nature 428:736-738.

Hewitt, C. D. 1998. A fully coupled GCM simulation of the climate of the mid-Holocene. Geophysical Research Letters 25:361-364.

Hodell, D. A., J. H. Curtis, G. A. Jones, A. Higuera-Gundy, M. Brenner, M. W. Binford, and K. T. Dorsey. 1991. Reconstruction of Caribbean climate change over the past 10,500 years. Nature 352:790-793.

Hodell, D. A., M. Brenner, J. H. Curtis, and T. Guilderson. 2001. Solar forcing of drought frequency in the Maya lowlands. Science 292:1367-1370.

Hoerling, M., and A. Kumar. 2003. The perfect ocean for drought. Science 299:691-694.

Holland, E. A., B. H. Braswell, J. Sulzman, and J-F. Lamarque. 2004. Nitrogen deposition onto the United States and Western Europe: A synthesis of observations and models. Ecological Applications (in press).

Hong, Y. T., B. Hong, Q. H. Lin, Y. X. Zhu, Y. Shibata, M. Hirota, M. Uchida, X. T. Leng, H. B. Jiang, H. Xu, H. Wang, and L. Yi. 2003. Correlation between Indian Ocean summer monsoon and North Atlantic climate during the Holocene. Earth and Planetary Science Letters 211:371-380.

Hood, L. L. 2003. Thermal response of the tropical tropopause region to solar ultraviolet variations. Geophysical Research Letters 30(23):2215, DOI: 10.1029/2003GL018364.

Houweling, S., T. Kaminski, F. Dentener, J. Lelieveld, and M. Heimann. 1999. Inverse modeling of methane sources and sinks using the adjoint of a global transport model. Journal of Geophysical Research 104:26137-26160.

Hu, F. S., D. Kaufman, S. Yoneji, D. Nelson, A. Semesh, S. Huang, J. Tian, G. Bond, B. Clegg, and T. Brown. 2003. Cyclic variation and solar forcing of Holocene climate in the Alaskan subarctic. Science 301:1890-1893.

Huebert, B. J., and R. J. Charlson. 2000. Uncertainties in data on organic aerosols. Tellus 52B:1249-1255.

Indermühle A., T. F. Stocker, F. Joos, H. Fischer, H. J. Smith, M. Wahlen, B. Seck, D. Mastrolanni, J. Tchumi, T. Blunier, R. Meyer, and B. Stauffer. 1999. Holocene carboncycle dynamics based on CO_2 trapped in the ice at Taylor Dome, Antarctica. Nature 398:121-126.

IPCC (Intergovernmental Panel on Climate Change). 1990. Scientific Assessment of Climate Change—Report of Working Group I, J. T. Houghton, G. J. Jenkins, and J. J. Ephraums, eds. Cambridge, U.K.: Cambridge University Press.

IPCC. 1992. Climate Change 1992—The Supplementary Report to the IPCC Scientific Assessment, J. T. Houghton, B. A. Callander, and S. K. Varney, eds. Cambridge, U.K.: Cambridge University Press.

IPCC. 1996. Climate Change 1995: The Science of Climate Change. Contribution of Working Group 1 to the Second Assessment Report of the Intergovernmental Panel on Climate Change, J. T. Houghton, L. G. Meira Filho, B. A. Callander, N. Harris, A. Kattenberg, and K. Maskell, eds. Cambridge, U. K.: Cambridge University Press.

IPCC. 2001. Climate Change 2001: The Scientific Basis. Contribution of Working Group I to the Third Assessment Report of the Intergovernmental Panel on Climate Change, J. T. Houghton, Y. Ding, D. J. Griggs, M. Noguer, P. J. van der Linden, X. Dai, K. Maskell, and C. A. Johnson, eds. Cambridge, U.K.: Cambridge University Press.

Jackman, C. H., E. L. Fleming, S. Chandra, D. B. Considine, and J. E. Rosenfield. 1996. Past, present and future modeled ozone trends with comparisons to observed trends. Journal of Geophysical Research 101:28753-28767.

Jackman, C. H., R. D. McPeters, G. J. Labow, E. L. Fleming, C. J. Praderas, and J. M. Russell. 2001. Northern Hemisphere atmospheric effects due to the July 2000 solar proton event. Geophysical Research Letters 28(15):2883-2886, DOI: 10.1029/2001GL013221.

Jacobson, M. Z. 2001. Global direct radiative forcing due to multicomponent anthropogenic and natural aerosols. Journal of Geophysical Research 106:1551-1568.

Jacobson, M. Z. 2002. Control of fossil-fuel particulate black carbon and organic matter, possibly the most effective method of slowing global warming. Journal of Geophysical Research 107(D19):4410, DOI: 10.1029/JD001376.

Jacobson, M. Z., R. P. Turco, E. J. Jensen, and O. B. Toon. 1994. Modeling coagulation among particles of different composition and size. Atmospheric Environment 28A:1327-1338.

Jirikowic, J. L., and P. E. Damon. 1994. The medieval solar activity maximum. Climatic Change 26:309-316.

Johnson, D. W., S. Osborne, R. Wood, K. Suhre, R. Johnson, S. Businger, P. K. Quinn, A. Wiedensohler, P. A. Durkee, L. M. Russell, M. O. Andreae, C. O'Dowd, K. Noone, B. Bandy, J. Rudolph, and S. Rapsomanikis. 2000. An overview of the Lagrangian experiments undertaken during the North Atlantic Regional Aerosol Characterization Experiment (ACE-2). Tellus 52B:290-320.

Jones, J. B., Jr., E. H. Stanley, and P. J. Mulholland. 2003. Long-term decline in carbon dioxide supersaturation in rivers across the contiguous United States. Geophysical Research Letters 30(10):1495, DOI:10.1029/2003GL017056.

Jones, P. D., and M. E. Mann. 2004. Climate over past millennia. Reviews of Geophysics 42:RG2002.

Jones, P. D., T. J. Osborn, K. R. Briffa, C. K. Folland, E. B. Horton, L. V. Alexander, D. E. Parker, and N. A. Rayner. 2001. Adjusting for sampling density in grid box land and ocean surface temperature time series. Journal of Geophysical Research 106:3371-3380.

Joshi, M., K. Shine, M. Ponater, N. Stuber, R. Sausen, and L. Li. 2003. A comparison of climate response to different radiative forcings in three general circulation models: Towards an improved metric of climate change. Climate Dynamics 20(7-8):843-854.

Kabat, P., M. Claussen, P. A. Dirmeyer, J. H. C. Gash, L. Bravo de Guenni, M. Meybeck, R. Pielke Sr., C. J. Vörösmarty, R. W. A. Hutjes, and S. Lütkemeier, eds. 2004. Vegetation, Water, Humans and the Climate—A New Perspective on an Interactive System. Berlin: Springer Verlag.

Kalnay, E., and M. Cai. 2003. Impact of urbanization and land-use change on climate. Nature 423:528-531.

Kalnay, E., M. Kanamitsu, R. Kistler, W. Collins, D. Deaven, L. Gandin, M. Iredell, S. Saha, G. White, J. Woolen, Y. Zhu, M. Chelliah, W. Ebisuzaki, W. Higgins, J. Janowiak, K. C. Mo, C. Ropelewski, J. Wang, A. Leetma, R. Reynolds, R. Jenne, and D. Joseph. 1996. The NCEP/NCAR 40-year reanalysis project. Bulletin of the American Meteorological Society 77:437-471.

Kaminski T., W. Knorr, P. J. Rayner, and M. Heimann. 2002. Assimilating atmospheric data into a terrestrial biosphere model: A case study of the seasonal cycle. Global Biogeochemical Cycles 16(4):1066, DOI: 10.1029/2001GB001463.

Karl, T. R., V. E. Derr, D. R. Easterling, C. K. Folland, D. J. Hoffman, S. Levitus, N. Nicholls, D. E. Parker, and G. W. Withee. 1995. Critical issues for long-term climate monitoring. Climatic Change 31:185-221.

Karoly, D. J., K. Braganza, P. A. Stott, J. M. Arblaster, G. A. Meehl, A. J. Broccoli, and K. W. Dixon. 2003. Detection of a human influence on North American climate. Science 302:1200-1203.

Kasibhatla, P., A. Arellano, J. A. Logan, P. I. Palmer, and P. Novelli. 2002. Top-down estimate of a large source of atmospheric carbon monoxide associated with fuel combustion in Asia. Geophysical Research Letters 29(19):1900, DOI: 10.1029/2002GL015581.

Kasting, J. F. 1993. Earth's early atmosphere. Science 259:920-928.

Kasting, J. F., and D. Catling. 2003. Evolution of a habitable planet. Annual Review of Astronomy and Astrophysics 41:429-463.

Katz, M. E., B. S. Cramer, G. S. Mountain, S. Katz, and K. G. Miller. 2001. Uncorking the bottle: What triggered the Paleocene/Eocene thermal maximum methane release? Paleoceanography 16:549-562.

Kaufman, Y. J., and R. S. Fraser. 1997. The effect of smoke particles on clouds and climate forcing. Science 277:1636-1639.

Kaufman, Y. J., D. Tanré, B. N. Holben, S. Mattoo, L. A. Remer, T. F. Eck, J. Vaughan, and B. Chatenet. 2002. Aerosol radiative impact on spectral solar flux at the surface, derived from principal plane sky measurements. Journal of the Atmospheric Sciences 59:635-646.

Kaya, Y. 1990. Impact of carbon dioxide emission control on GNP growth: Interpretation of proposed scenarios. Paper presented to the IPCC Energy and Industry Subgroup, Response Strategies Working Group, Paris.

Keigwin, L. D., and E. A. Boyle. 2000. Detecting Holocene changes in thermohaline circulation. Proceedings of the National Academy of Sciences 97(4):1343-1346.

Kellogg, W. W., and S. H. Schneider. 1974. Climate stabilization: For better or for worse? Science 186:1163-1172.

Kent, D. V., B. S. Cramer, L. Lance, D. Wang, J. D. Wright, and R. Van der Voo. 2003. A case for a comet impact trigger for the Paleocene/Eocene thermal maximum and carbon isotope excursion. Earth and Planetary Science Letters 211:13-26.

Khain, A. P., D. Rosenfeld, and A. Pokrovsky, A. 2001. Simulating convective clouds with sustained supercooled liquid water down to $-37.5°C$ using a spectral microphysics model. Geophysical Research Letters 28:3887-3890.

Khain, A., A. Pokrovsky, M. Punsky, A. Seifert, and V. Phillips. 2004. Simulation of effects of atmospheric aerosols on deep turbulent convective clouds using a spectral microphysics mixed-phase cumulus cloud model. Part 1: Model description and possible applications. Journal of Atmospheric Sciences. 61(24):2963-2983.

Khalil, M. A. K., and M. J. Shearer. 1993. Sources of methane: An overview. In Atmospheric Methane, Sources, Sinks and Role in Global Change, M. A. K. Khalil, ed. Berlin: Springer Verlag.

Kiehl, J. T., and B. P. Briegleb. 1993. The relative roles of sulfate aerosols and greenhouse gases in climate forcing. Science 260:311-314.

Kiehl, J. T., and V. Ramanathan. 1982. Radiative heating due to increased CO_2: The role of H_2O continuum absorption. Journal of the Atmospheric Sciences, 39: 2923-2926.

Kiehl, J. T., and K. E. Trenberth, 1997. Earth's annual mean energy budget. Bulletin of the American Meteorological Society 78:197-208.

Kiehl, J. T., J. J. Hack, M. H. Zhang, and R. D. Cess. 1995. Sensitivity of a GCM climate to enhanced shortwave cloud absorption. Journal of Climate 8:2200-2212.

Kiehl, J. T., J. J. Hack, and J. W. Hurrell. 1998. The energy budget of the NCAR community climate model CCM3. Journal of Climate 11:1151-1178.

Kim, B. G., S. E. Schwartz, M. A. Miller, and Q. L. Min. 2003. Effective radius of cloud droplets by ground-based remote sensing: Relationship to aerosol. Journal of Geophysical Research-Atmospheres. 108(D23), DOI: 10.1029/2003JD0003721.

Kirchner, I., G.L. Stenchikov, H.-F. Graf, A. Robock, and J.C. Antuna. 1999. Climate model simulation of winter warming and summer cooling following the 1991 Mount Pinatubo volcanic eruption, Journal of Geophysical Research 104(D16):19039-19055.

Kitoh, A., and S. Murakami. 2002. Tropical Pacific climate at the mid-Holocene and the Last Glacial Maximum simulated by a coupled ocean-atmosphere general circulation model. Paleoceanography 17:1-13.

Klein Goldewijk, K. 2001. Estimating global land use change over the past 300 years: The HYDE database. Global Biogeochemical Cycles 15(2):417-434.

Knaff, J. A., and C. W. Landsea. 1997. An El Niño-Southern Oscillation CLImatology and PERsistence (CLIPER) Forecasting Scheme. Weather and Forecasting 12:633-652.

Knutti, R., T. F. Stocker, F. Joos, and G. K. Plattner. 2002. Constraints on radiative forcing and future climate change from observations and climate model ensembles. Nature 416(6882):719-723.

Koch, D. 2001. Transport and direct radiative forcing of carbonaceous and sulfate aerosols in the GISS GCM. Journal of Geophysical Research 106(D17):20311-20332, DOI: 10.1029/2001JD900038.

Koch, D., D. Jacob, I. Tegen, D. Rind, and M. Chin. 1999. Tropospheric sulfur simulation and sulfate direct radiative forcing in the Goddard Institute for Space Studies general circulation model. Journal of Geophysical Research 104:23799-23822.

Kodera K. 2002. Solar cycle modulation of the North Atlantic Oscillation: Implication in the spatial structure of the NAO. Geophysical Research Letters 29(8), DOI: 10.1029/2001GL014557.

Koren, I., Y. J. Kaufman, L. A. Remer, and J. V. Martins. 2004. Measurement of the effect of Amazon smoke on inhibition of cloud formation. Science 303:1342-1345.

Krakauer, N. Y., and J. T. Randerson. 2003. Do volcanic eruptions enhance or diminish net primary production? Evidence from tree rings. Biogeochemical Cycles 17(4):1118, DOI: 10.1029/2003GB002076.

Kristjánsson, J. E. 2002. Studies of the aerosol indirect effect from sulfate and black carbon aerosols, Journal of Geophysical Research 107(D15):4246, DOI: 10.1029/2001JD000887.

Kristjánsson, J. E., A. Staple, and J. Kristiansen. 2002. A new look at possible connections between solar activity, clouds and climate, Geophysical Research Letters 29(23):2107, DOI: 10.1029/2002GL015646.

Krüger O., and H. Graßl. 2002. The indirect aerosol effect over Europe. Geophysical Research Letters 29(19)311-314, DOI: 10.1029/2001GL014081.

Krüger, O., and H. Graßl. 2004. Albedo reduction by absorbing aerosols over China. Geophysical Research Letters 31(2):L02108, DOI: 10.1029/2003GL019111.

Kukla, G., and J. Gavin. 2004. Milankovitch climate reinforcements. Global and Planetary Change 40:27-48.

Kump, L. R., S. L. Brantley, and M. A. Arthur. 2000. Chemical weathering, atmospheric CO_2, and climate. Annual Review of Earth and Planetary Sciences 28:611-667.

Kvenvolden, K. A. 2002. Methane hydrate in the global organic carbon cycle. Terra Nova 14:302-306.

Kyoto Protocol to the United Nations Framework Convention on Climate Change. 1997. Document FCCC/CP/1997/L.7/Add.1. Available at *http://unfccc.int/resource/docs/convkp/kpeng.html*. Accessed December 2, 2004.

Labitzke, K., and K. Matthes. 2003. Eleven-year solar cycle variations in the atmosphere: Observations, mechanisms and models. Holocene 13:311-317.

Lacis, A. A., D. J. Wuebbles, and J. A. Logan. 1990. Radiative forcing by changes in the vertical distribution of ozone. Journal of Geophysical Research 95:9971-9981.

Lal, D. 1988. Theoretically expected variations in the terrestrial cosmic-ray production rates of isotopes. Pp. 216-233 in Solar-Terrestrial Relationships and the Earth Environment in the Last Millennia, G. C. Castagnoli, ed. Amsterdam: North-Holland Physics Publishing.

Lambin, E. F., H. J. Geist, and E. Lepers. 2003. Dynamics of land-use and land-cover change in tropical regions. Annual Review of Environment and Resources 28:205-241.

Lashof, D. A. 2000. The use of global warming potentials in the Kyoto Protocol. Climatic Change 44:423-425.

Laskar, J. 1990. The chaotic motion of the solar system: A numerical estimate of the size of the chaotic zones. Icarus 88:266-291.

Laurance, W. F., M. A. Cochrane, S. Bergen, P. M. Fearnside, P. Delamônica, C. Barber, S. D'Angelo, and T. Fernandes. 2001. The future of the Brazilian Amazon. Science 291(5503):438-439, DOI: 10.1126/science.291.5503.438

Lavorel, S., M. D. Flannigan, E. F. Lambin, and M. Scholes. 2005. Regional vulnerability to fire: Feedbacks, nonlinearities, and interactions. Bioscience (submitted).

Lawrence, M. G., P. Joeckel, and R. von Kuhlmann. 2001. What does the global mean OH concentrations tell us? Atmospheric Chemistry and Physics 1:37-49.

Lean, J. L. 2000. Evolution of the Sun's spectral irradiance since the Maunder Minimum. Geophysical Research Letters 27:2425-2428.

Lean, J. L. 2001. Solar irradiance and climate forcing in the near future. Geophysical Research Letters 28:4119-4122.

Lean, J., J. Beer, and R. Bradley. 1995. Reconstruction of solar irradiance since 1610: Implications for climate change. Geophysical Research Letters 22:3195-3198.

Lean, J. L., G. J. Rottman, H. L. Kyle, T. N. Woods, J. R. Hickey, and L. C. Puga. 1997. Detection and parameterization of variations in solar mid and near ultraviolet radiation (200 to 400 nm). Journal of Geophysical Research 102:29939-29956.

Lean, J., Y.-M. Wang, and N. R. Sheeley, Jr. 2002. The effect of increasing solar activity on the Sun's total and open magnetic flux during multiple cycles: Implications for solar forcing of climate. Geophysical Research Letters 29, DOI: 10.1029/2002GL015880.

Leggett, J., W. J. Pepper, and R. J. Swart. 1992. Emissions Scenarios for IPCC: An Update. Pp. 69-95 in Climate Change 1992. The Supplementary Report to the IPCC Scientific Assessment, J. T. Houghton, B. A. Callander, and S. K. Varney, eds. Cambridge, U.K.: Cambridge University Press.

Lelieveld, J., and P. J. Crutzen. 1992. Indirect chemical effects of methane on climate warming. Nature 355:339-342.

Lelieveld, J., P. J. Crutzen, and C. Brühl. 1993. Climate effects of atmospheric methane. Chemosphere 26:739-768.

Lelieveld, J., P. Crutzen, and F. J. Dentener. 1998. Changing concentration, lifetime and climate forcing of atmospheric methane. Tellus 50B:128-150.

Levitus, S. J. I. Antonov, T. P. Boyer, and C. Stephens, 2000. Warming of the world ocean. Science 287:2225-2229.

Levitus, S. J. I. Antonov, J. Wang, T. L. Delworth, K. W. Dixon, and A. J. Broccoli, 2001. Anthropogenic warming of earth's climate system. Science 292:267-270.

Li, Q. B., D. J. Jacob, R. Park, Y. X. Wang, C. L. Heald, R. Hudman, R. M. Yantosca, R. V. Martin, and M. J. Evans. 2004. Outflow pathways for North American pollution in summer: A global 3-D model analysis of MODIS and MOPITT observations. Journal of Geophysical Research (submitted).

Liepert, B. G. 2002. Observed reductions in surface solar radiation in the United States and worldwide from 1961 to 1990. Geophysical Research Letters 29(12), DOI: 10.1029/2002GL014910.

Liepert, B., J. Feichter, U. Lohmann, and E. Roeckner. 2004. Can aerosols spin down the water cycle in a warmer and moister world? Geophysical Research Letters 31, DOI: 10.1029/2003GL019060.

Liston, G. E., J. P. McFadden, M. Sturm, and R. A. Pielke, Sr. 2002. Modelled changes in arctic tundra snow, energy and moisture fluxes due to increased shrubs. Global Change Biology 8:17-32.

Liu, H. S., R. Kolenkiewicz, and C. Wade. 2003. Insolation pulsation theory for atmospheric greenhouse gas concentrations. Recent Research Developments in Atmospheric Sciences 3:57-79.

Liu, Y., and P. H. Daum. 2002. Indirect warming effect from dispersion forcing. Nature 419:580-581.

Logan, J. A. 1999. An analysis of ozonesonde data for the troposphere: Recommendations for testing 3-D models, and development of a gridded climatology for tropospheric ozone. Journal of Geophysical Research 104:16115-16149.

Lohmann, U. 2002. A glaciation indirect aerosol effect caused by soot aerosols. Geophysical Research Letters 29(4) DOI: 10.1029/2001GL014357.

Lohmann, U., and J. Feichter. 2001. Can the direct and semi-direct aerosol effect compete with the indirect effect on a global scale? Geophysical Research Letters 28:159-161.

Lohmann, U., and J. Feichter. 2004. Global indirect aerosol effects: A review. Atmospheric Chemistry and Physics Discussions 4:7561-7614,

Lohmann, U., and B. Kärcher. 2002. First interactive simulations of cirrus clouds formed by homogeneous freezing in the ECHAM GCM. Journal of Geophysical Research 107:4105, DOI: 10.1029/2001JD000767.

Lohmann, U., J. Feichter, C. C. Chuang, and J. E. Penner. 1999. Predicting the number of cloud droplets in the ECHAM GCM. Journal of Geophysical Research 104:9169-9198.

Lohmann, U., J. Feichter, J. E. Penner, and W. R. Leaitch. 2000. Indirect effect of sulfate and carbonaceous aerosols: A mechanistic treatment. Journal of Geophysical Research 105:12193-12206.

Loveland, T. R., B. C. Reed, J. F. Brown, D. O. Ohlen, J. Zhu, L. Yang, and J. W. Merchant. 2000. Development of a global land cover characteristics database and IGBP DISCover from 1-km AVHRR Data. International Journal of Remote Sensing 21(6/7):1303-1330.

Loveland, T. R., T. L. Sohl, S. V. Stehman, A. L. Gallant, K. L. Sayler, and D. E. Napton. 2002. A strategy for estimating the rates of recent United States land-cover changes. Photogrammetric Engineering and Remote Sensing 68(10):1091-1099.

Lutz, W. 1996. The Future Population of the World: What Can We Assume Today? Revised Edition. London, U.K.: Earthscan.

Lutz, W., W. C. Sanderson, and S. Scherbov. 2004. The End of World Population Growth in the 21st Century. London, U.K.: Earthscan.

Maddison, A. 1995. Monitoring the World Economy 1820-1992. OECD Development Centre Studies. Paris: Organisation for Economic Co-operation and Development.

Magny, M., and C. Bégeot. 2004. Hydrological changes in the European midlatitudes associated with freshwater outbursts from Lake Agassiz during the Younger Dryas event and the early Holocene. Quaternary Research 61(2):181-192.

Mahowald, N. M., and C. Luo. 2003. A less dusty future? Geophysical Research Letters 30(17):1903, DOI: 10.1029/2003GL017880.

Mahowald, N. M., R. G. Prinn, and P. J. Rasch. 1997. Deducing CCl_3F emissions using an inverse method and chemical transport models with assimilated winds. Journal of Geophysical Research 102:28153-28168.

Manabe, S., and A. J. Broccoli. 1985. The influence of continental ice sheets on the climate of an ice age. Journal of Geophysical Research 90(D1):2167-2190.

Manabe, S., and R. F. Strickler. 1964. Thermal equilibrium of the atmosphere with a convective adjustment. Journal of the Atmospheric Sciences 21(4):361-385.

Manabe, S., and R. Wetherald. 1967. Thermal equilibrium of the atmosphere with a given distribution of relative humidity. Journal of Atmospheric Science 24:241-259.

Manabe, S., and Wetherald, R. T. 1975. The effects of doubling the CO_2-concentration on the climate of a general circulation model. Journal of Atmospheric Science 32:3-15.

Mann, M. E., and P. D. Jones. 2003. Global surface temperatures over the past two millennia. Geophysical Research Letters 30:1820, DOI: 10.1029/2003GL017814.

Mann, M. E., R. S. Bradley, and M. K. Hughes. 1998. Global-scale temperature patterns and climate forcing over the past six centuries. Nature 392:779-787.

Mann, M. E., C. M. Ammann, R. S. Bradley, K. R. Briffa, T. J. Crowley, M. K. Hughes, P. D. Jones, M. Oppenheimer, T. J. Osborn, J. T. Overpeck, S. Rutherford, K. E. Trenberth, and T. M. L. Wigley. 2003. On past temperatures and anomalous late 20th century warmth. Eos 84:256-258.

Mann, M. E., M. A. Cane, S. E. Zebiak, and A. Clement. 2005. Volcanic and solar forcing of the tropical Pacific over the past 1000 years. Journal of Climate 18:447-456.

Manne, A. S., and R. G. Richels. 2001. An alternative approach to establishing trade-offs among greenhouse gases. Nature 410(6829):675-677.

Manne, A.S., R. Mendelsohn, and R. Richels. 1995. MERGE—A model for evaluating regional and global effects of GHG reduction policies. Energy Policy 23(1):17-34.

Marland, G., R. A. Pielke, Sr., M. Apps, R. Avissar, R. A. Betts, K. J. Davis, P. C. Frumhoff, S. T. Jackson, L. Joyce, P. Kauppi, J. Katzenberger, K. G. MacDicken, R. Neilson, J. O. Niles, D. S. Niyogi, R. J. Norby, N. Peña, N. Sampson, and Y. Xue. 2003. The climatic impacts of land surface change and carbon management, and the implications for climate-change mitigation policy. Climate Policy 3:149-157.

Marley, N., J. Gaffney, C. Baird, C. Blazer, P. Drayton, and J. Frederick. 2001. An empirical method for the determination of the complex refractive index of size-fractionated atmospheric aerosols for radiative transfer calculations. Aerosol Science and Technology: 34:535-549.

Marotzke, J. 2000. Abrupt climate change and the thermohaline circulation: Mechanisms and predictability. Proceedings of the National Academy of Sciences 97:1347-1350.

Marotzke, J., and B. A. Klinger. 2000. The dynamics of equatorially asymmetric thermohaline circulations. Journal of Physical Oceanography 30(5):955-970.

Marshall, C. H., R. A. Pielke Sr., and L. T. Steyaert. 2003. Wetlands: Crop freezes and land-use change in Florida. Nature 426:29-30.

Marshall, C. H., R. A. Pielke Sr., and L. T. Steyeart. 2004a. Has the conversion of natural wetlands in agricultural land increased the incidence and severity of damaging freezes in south Florida? Monthly Weather Review 132(9):2243–2258, DOI: 10.1175/1520-0493(2004)132.

Marshall, C. H., R. A. Pielke Sr., L. T. Steyaert, and D. A. Willard. 2004b. The impact of anthropogenic land-cover change on the Florida peninsula sea breezes and warm season sensible weather. Monthly Weather Review 132:28-52.

Martin, R. V., K. Chance, D. J. Jacob, T. P. Kurosu, R. J. D. Spurr, E. Bucsela, J. F. Gleason, P. I. Palmer, I. Bey, A. M. Fiore, Q. Li, R. M. Yantosca, and R. B. A. Koelemeijer. 2002. An improved retrieval of tropospheric nitrogen dioxide from GOME. Journal of Geophysical Research 107(D20):4437, DOI: 10.1029/2001JD001027, 2002.

Martin, S. T., H. M. Hung, R. J. Park, D. J. Jacob, R. J. D. Spurr, K. V. Chance, and M. Chin. 2004. Effects of the physical state of tropospheric ammonium-sulfate-nitrate particles on global aerosol direct radiative forcing. Atmospheric Chemistry and Physics 4:183-214.

Matsui, T., H. Masunaga, R. A. Pielke Sr., and W-K. Tao. 2004. Impact of aerosols and atmospheric thermodynamics on cloud properties within the climate system. Geophysical Research Letters 31, L06109, DOI:10.1029/2003GL019287.

Matsuoka, Y., T. Morita, and M. Kainuma. 2001. Integrated assessment model of climate change: The AIM Approach. Pp. 339-361 in Present and Future of Modeling Global Environmental Change. Tokyo, Japan: Terra Scientific Publishing Company.

Matthes, F. 1939. Report of Committee on Glaciers. Transactions, American Geophysical Union 20:518-523.

Matthews, E. 1983. Global vegetation and land use: New high-resolution data bases for climate studies. Journal of Climate and Applied Meteorology 22:474-487.

McCormack, J. P., L. L. Hood, and R. D. McPeters. 1997. Approximate separation of volcanic and 11-year signals in the SBUV-SBUV/2 total ozone record over the 1979-1995 period. Geophysical Research Letters 24:2729.

McFadden, J. P., G. E. Liston, M. Sturm, R. A. Pielke Sr., and F. S. Chapin III. 2001. Interactions of shrubs and snow in Arctic tundra: Measurements and models. Pp. 317-325 in Soil-Vegetation-Atmosphere Transfer Schemes and Large-Scale Hydrological Models, A. J. Dolman, A. J. Hall, M. L. Kavvas, T. Oki, and J. W. Pomeroy, eds. International Association of Hydrological Sciences Publication No. 270. Wallingford, Oxfordshire, U.K.: IAHS Press.

McFarquhar, G., and A. Heymsfield. 2001. Parameterizations of INDOEX microphysical measurements and calculations of cloud susceptibility: Applications for climate studies. Journal of Geophysical Research 106(D22):28675-28698.

McGraw, R. 1997. Description of aerosol dynamics by the quadrature method of moments. Aerosol Science and Technology 27:255-265.

Mears, C. A., M. Schabel, and F. J. Wentz. 2003. A reanalysis of the MSU Channel 2 tropospheric temperature record. Journal of Climate 16:3650-3664.

Menon, S., A. D. Del Genio, D. Koch, and G. Tselioudis. 2002a. GCM simulations of the aerosol indirect effect: sensitivity to cloud parameterization and aerosol burden, Journal of Atmospheric Science 59:692-713.

Menon, S., J. E. Hansen, L. Nazarenko, and Y. Luo. 2002b. Climate effects of black carbon aerosols in China and India. Science 297:2250-2253.

Messner, S., and M. Strubegger. 1995. User's Guide for MESSAGE III, WP-95-69. Laxenburg, Austria: International Institute for Applied Systems Analysis.

Mickley, L. J., P. P. Murti, D. J. Jacob, J. A. Logan, D. Rind, and D. Koch. 1999. Radiative forcing from tropospheric ozone calculated with a unified chemistry-climate model. Journal of Geophysical Research 104:30153-30172.

Mickley, L. J., D. J. Jacob, and D. Rind. 2001. Uncertainty in preindustrial abundance of tropospheric ozone: Implications for radiative forcing calculations. Journal of Geophysical Research 106:3389-3399.

Mickley, L. J., D. J. Jacob, B. D. Field, and D. Rind. 2004. Climate response to the increase in tropospheric ozone since preindustrial times: A comparison between ozone and equivalent CO_2 forcings. Journal of Geophysical Research 109:D05106, DOI: 1029/2003JD003653.

Milkov, A. V. 2004. Global estimates of hydrate-bound gas in marine sediments: How much is really out there? Earth Sciences Review 66(3-4):183-197.

Ming, Y., and L. M. Russell. 2002. Thermodynamic equilibrium of organic-electrolyte mixtures in aerosol particles. American Institute of Chemical Engineers Journal 48:1331-1348.

Ming, Y., and L. M. Russell. 2004. Organic aerosol effects on fog droplet spectra. Journal of Geophysical Research 109(D10):206, DOI: 10.1029/2003JD004427.

Moller, F. 1963. On the influence of changes in CO_2 concentration in air on the radiation balance of Earth's surface and on climate. Journal of Geophysical Research 68:3877-3886.

Monson, R. K., and R. Fall. 1989. Isoprene emission from aspen leaves. Plant Physiology 90:267-274.

Morgan, J. P., T. J. Reston, and C. R. Ranero. 2004. Contemporaneous mass extinctions, continental flood basalts, and "impact signals": Are mantle plume-induced lithospheric gas explosions the causal link? Earth and Planetary Science Letters 217:263-284.

Mori, S., and M. Takahashi. 1999. An integrated assessment model for the evaluation of new energy technologies and food productivity. International Journal of Global Energy Issues 11(1-4):1-18.

Morita, T., Y. Matsuoka, I. Penna, and M. Kainuma. 1994. Global Carbon Dioxide Emission Scenarios and Their Basic Assumptions: 1994 Survey. CGER-1011-94. Tsukuba, Japan: Center for Global Environmental Research, National Institute for Environmental Studies.

Morrical, B. D., D. P. Fergenson, and K. A. Prather. 1998. Coupling two-step laser desorption/ionization with aerosol time-of-flight mass spectrometry for the analysis of individual organic aerosol particles. Journal of the American Society of Mass Spectrometry 9:1068-1073.

Moy, C. M., G. O. Seltzer, D. T. Rodbell, and D. M. Anderson. 2002. Variability of El Niño/Southern Oscillation activity at millennial timescales during the Holocene epoch. Nature 420:162-165.

Murphy, J. M. 1995. Transient response of the Hadley Centre coupled ocean-atmosphere model to increasing carbon dioxide. Part I: Control climate and flux adjustment. Journal of Climate 8:36-56.

Murphy, D. M., D. S. Thomson, M. Kaluzhny, J. J. Marti, and R. J. Weber. 1997. Aerosol characteristics at Idaho Hill during the OH Photochemistry Experiment. Journal of Geophysical Research 102(D5):6325-6330, DOI: 10.1029/96JD02552.

Myhre, G., and A. Myhre. 2003. Uncertainties in radiative forcing due to surface albedo changes caused by land-use changes. Journal of Climate 16:1511-1524.

Myhre, G., and F. Stordal. 1997. Role of spatial and temporal variations in the computation of radiative forcing and GWP. Journal of Geophysical Research 102:11181-11200.

Nakajima, T., A. Higurashi, K. Kawamoto, and J. E. Penner. 2001. A possible correlation between satellite-derived cloud and aerosol microphysical parameters. Geophysical Research Letters 28:1171-1174.

Nakićenović, N. 1996. Freeing energy from carbon. Daedalus 125(3):95-112.

Nakićenović, N. A. Grübler, and A. McDonald (eds.) 1998. P. 299 in Global Energy Perspectives. Cambridge, U.K.: Cambridge University Press.

Nakićenović, N., J. Alcamo, G. Davis, B. deVries, J. Fenhann, S. Gaffin, K. Gregory, A. Grübler, T. Y. Jung, T. Kram, E. Lebre LaRovere, L. Michaelis, S. Mori, T. Morita, W. Pepper, H. Pitcher, L. Price, K. Riahi, A. Roehrl, H.-H. Rogner, A. Sankovski, M. Schlesinger, P. Shukla, S. Smith, R. Swart, S. van Rooijen, N. Victor, and Z. Dadi. 2000. IPCC Special Report on Emissions Scenarios. Available online at *http://www.grida.no/climate/ipcc/emission/*, accessed July 15, 2004.

Napton, D. E., T. L. Sohl, R. F. Auch, and T. R. Loveland. 2003. Land use and land cover change in the north central Appalachians Ecoregion. Pennsylvania Geographer XLI(1):1-22 (in press).

Narisma, G. T., A. J. Pitman, J. Eastman, I. G. Watterson, R. Pielke Sr., and A. Beltran-Przekurat. 2003. The role of biospheric feedbacks in the simulation of the impact of historical land cover change on the Australian January climate. Geophysical Research Letters 30(22):2168, DOI: 10.1029/2003GL018261.

NARSTO (North American Research Strategy for Tropospheric Ozone). 2003. Particulate Matter Science for Policy Makers: A NARSTO Assessment. Electric Power Research Institute (EPRI) Report 1007735. Palo Alto, Calif.: EPRI.

NAST (National Assessment Synthesis Team). 2001. Climate Change Impacts on the United States: The Potential Consequences of Climate Variability and Change. Report for the U.S. Global Change Research Program. Cambridge, U.K.: Cambridge University Press.

Neff, U., S. J. Burns, A. Mangini, M. Mudelsee, D. Fleitmann, and A. Matter. 2001. Strong coherence between solar variability and the monsoon in Oman between 9 and 6 kyr ago. Nature 411:290-293.

Nenes, A., and J. H. Seinfeld. 2003. Parameterization of cloud droplet formation in global climate models. Journal of Geophysical Research 108(D14):4415, DOI: 10.1029/2002JD002911.

Nerem, R. S., and G. T. Mitchum. 2001. Observation of sea level change from satellite altimetry. In Sea Level Rise, B. C. Douglas, M. S. Kearney, and S. P. Leartherman, eds. San Diego, Calif.: Academic Press.

Nesbitt, S. W., R. Zhang, and R. E. Orville. 2000. Seasonal and global NO_x production by lightning estimated from the Optical Transient Detector (OTD). Tellus B 52:1206-1215.

Niggemann, S., A. Mangini, M. Mudelsee, D. K. Richter, and G. Wurth. 2003. Sub-Milankovitch climate cycles in Holocene stalagmites from Sauerland, Germany. Earth and Planetary Science Letters 216(4):539-547.

Niyogi, D., H. I. Chang, V. K. Sexana, T. Holt, K. Alapaty, F. L. Booker, F. Chen,K. J. Davies, B. Holben, T. Matsui, T. Meyers, W. C. Oechel, R. A. Pielke, R. Wells, K. Wilson, and X. Yongkang. 2004. Direct observations of the effects of aerosol loading on net ecosystem CO_2 exchanges over different landscapes. Geophysical Research Letters 31(l20506), DOI: 10.1029/2004GL020915.

Nober, F. J., H.-F. Graf, and D. Rosenfeld. 2003. Sensitivity of the global circulation to the suppression of precipitation by anthropogenic aerosols. Global Planetary Change 37:57-80.

Noone, K. J., J. A. Ogren, A. Hallberg, J. Heintzenberg, J. Ström, H. C. Hansson, B. Svenningsson, A. Wiedensohler, S. Fuzzi, M. C. Facchini, B. G. Arends, and A. Berner. 1992. Changes in aerosol size and phase distributions due to physical and chemical processes in fog. Tellus 44B(5):489-504.

Nordhaus, W. D. 1993. Optimal greenhouse-gas reductions and tax policy in the DICE model. AEA Papers and Proceedings 83(2).

Nordhaus, W. D., and J. Boyer. 2000. Warming the World: Economic Models of Global Warming. Cambridge, Mass.: MIT Press.

Noren, A. J., P. R. Bierman, E. J. Steig, A. Lini, and J. Southon. 2002. Millennial-scale storminess variability in the northeastern Unites States during the Holocene epoch. Nature 419:821-824.

Novakov, T., V. Ramanathan, J. E. Hansen, T. W. Kirchstetter, Mki. Sato, J. E. Sinton, and J. A. Satahye. 2003. Large historical changes of fossil-fuel black carbon aerosols. Geophysical Research Letters 30(6):1324, DOI: 10.1029/2002GL016345.

Noyes, R. W. 1982. The Sun, Our Star. Cambridge, Mass.: Harvard University Press.

NRC (National Research Council). 1979. Carbon Dioxide and Climate: A Scientific Assessment. Washington, D.C.: National Academy Press.

NRC. 1988. Toward an Understanding of Global Change: Initial Priorities for U.S. Contributions to the International Geosphere–Biosphere Program. Washington, D.C.: National Academy Press.

NRC. 1999. Adequacy of Climate Observing Systems. Washington, D.C.: National Academy Press.

NRC. 2000. Reconciling Observations of Global Temperature Change. Washington, D.C.: National Academy Press.

NRC. 2001. Climate Change Science: An Analysis of Some Key Questions. Washington, D.C.: National Academy Press.

NRC. 2002. Abrupt Climate Change: Inevitable Surprises. Washington, D.C.: National Academy Press.

NRC. 2003. Understanding Climate Change Feedbacks. Washington, D.C.: The National Academies Press.

NRC. 2004. Air Quality Management in the United States. Washington, D.C.: The National Academies Press.

O'Brien, J. J. 2001. Ph.D. thesis, The effects of climate on the growth and physiology of tropical rainforest canopy trees. Florida International University, Miami.

Ogren, J. A., and R. J. Charlson. 1983. Elemental carbon in the atmosphere: Cycle and lifetime. Tellus 35B:241-254.

Ohmura, A., E. Dutton, B. Forgan, C. Fröhlich, H. Gilgen, H. Hegner, A. Heimo, G. König-Langlo, B. McArthur, G. Müller, R. Philipona, P. Pinker, C. H. Whitlock, K. Dehne, and M. Wild. 1998. Baseline Surface Radiation Network (BSRN/WCRP), a new precision radiometry for climate research. Bulletin of the American Meteorological Society 79:2115-2136.

Ohring, G., B. Wielicki, R. Spencer, W. Emery, and R. V. Datla, eds. 2004. Satellite Instrument Calibration for Measuring Global Climate Change. U. S. Department of Commerce. National Institute of Standards and Technology. NISTIR 7047.

Olivier, J. G. J, A. F. Bouwman, C. W. M. van der Maas, J. J. M. Berdowski, C. Veldt, J. P. J. Bloos, A. J. H. Visschedijk, P. Y. J. Zanfeld, and J. L. Haverlag. 1996. Description of EDGAR Version 2.0: A Set of Global Emission Inventories of Greenhouse Gas Gases and Ozone-Depleting Substances for all Anthropogenic and Most Natural Sources on a per Country Basis and on a 1 × 1 grid. Rijksinstituut voor Volksgezondheid en Milieu (RIVM) Report 771060 002. Bilthoven, The Netherlands: RIVM.

O'Neill, B. C. 2000. The jury is still out on global warming potentials. Climatic Change 44:427-443.

Pak, B. C., and M. J. Prather. 2001. CO_2 source inversions using satellite observations of the upper troposphere. Geophysical Research Letters 28:4571-4574.

Palmer, P. I., D. J. Jacob, L. J. Mickley, D. R. Blake, G. W. Sachse, H. E. Fuelberg, and C. M. Kiley. 2003. Eastern Asian emissions of anthropogenic halocarbons deduced from aircraft concentration data. Journal of Geophysical Research 108(D24):4753, DOI: 10.1029/2003JD00359.

Pandis, S. N., L. M. Russell, and J. H. Seinfeld. 1994. The relationship between the DMS flux and the CCN concentration in remote marine regions. Journal of Geophysical Research 99:16945-16957.

Pawlowska, H., and J-L Brenguier. 2000. Microphysical properties of stratocumulus clouds during ACE-2. Tellus 52(2):868-887.

Peng, Y., and U. Lohmann. 2003. Sensitivity study of the spectral dispersion of the cloud droplet size distribution on the indirect aerosol effect. Geophysical Research Letters 30, DOI: 10.1029/2003GL017192.

Penner, J. E., D. H. Lister, D. J. Griggs, D. Docken, and M. MacFarland, eds. 1999. Aviation and the Global Atmosphere, Intergovernmental Panel on Climate Change Special Report. Cambridge, U.K.: Cambridge University Press.

Penner, J. E., S. Y. Zhang, and C. C. Chuang. 2003. Soot and smoke aerosol may not warm climate. Journal of Geophysical Research 108, DOI: 10.1029/2003JD003409

Pepper, W. J., W. Barbour, A. Sankovski, and B. Braaz. 1998. No-policy greenhouse gas emission scenarios: Revisiting IPCC 1992. Environmental Science and Policy 1:289-312.

Perry, C. A., and K. J. Hsu. 2000. Geophysical, archaeological, and historical evidence support a solar-output model for climate change. Proceedings of the National Academy of Sciences 97:12433-12438.

Peterson, L. C., G. H. Haug, K. A. Hughen, and U. Rohl. 2000. Rapid changes in the hydrologic cycle of the tropical Atlantic during the last glacial. Science 290:1947-1951.

Petron, G., C. Grainer, B. Khattatov, J. F. Lamarque, V. Yudin, J. F.Müller, and J. Gille. 2002. Inverse modeling of carbon monoxide surface emissions using CMDL network observations. Journal of Geophysical Research 107(D24):4761, DOI:10.1029/2001JD001305.

Peylin, P., D. Baker, J. Sarmiento, P. Ciais, P. Bousquet. 2002. Influence of transport uncertainty on annual mean and seasonal inversions of atmospheric CO_2 data. Journal of Geophysical Research 107:D19(4385), DOI: 10.1029/2001JD000857.

Pielke, R. A., Jr. 2001. Room for doubt. Nature 410:151.

Pielke, R. A., Sr. 2003. Heat storage within the Earth system. Bulletin of the American Meteorological Society 84:331-335

Pielke, R. A., Sr., and T. N. Chase. 2003. A Proposed New Metric for Quantifying the Climatic Effects of Human-Caused Alterations to the Global Water Cycle. Presented at the Symposium on Observing and Understanding the Variability of Water in Weather and Climate, 83rd American Meteorological Society Annual Meeting, Long Beach, Calif., February 9-13, 2003.

Pielke, R. A., Sr., and T. N. Chase. 2004. Technical comment: Contributions of anthropogenic and natural forcing to recent tropopause height changes. Science 303:1771b.

Pielke, R. A., Sr., T. N. Chase, T. G. F. Kittel, J. Knaff, and J. Eastman. 2001. Analysis of 200 mbar zonal wind for the period 1958-1997. Journal of Geophysical Research 106(D21):27287-27290.

Pielke R. A., Sr., G. Marland, R. A. Betts, T. N. Chase, J. L. Eastman, J. O. Niles, D. Niyogi, and S. Running. 2002. The influence of land-use change and landscape dynamics on the climate system—Relevance to climate change policy beyond the radiative effect of greenhouse gases. Philosophical Transactions of the Royal Society of London Series A 360:1705-1719.

Piexoto, J. P. and A. H. Oort, 1992: Physics of Climate. New York: American Institute of Physics, 520pp.

Pitman, A. J. 2003. Review: The evolution of, and revolution in, land surface schemes designed for climate models. International Journal of Climatology 23:479-510.

Platnick, S., P. A. Durkee, K. Nielsen, J. P. Taylor, S. C. Tsay, M. D. King, R. J. Ferek, P. V. Hobbs, and J. W. Rottman. 2000. The role of background cloud microphysics in the radiative formation of ship tracks. Journal of the Atmospheric Sciences 57(16):2607-2624.

Plutzar, C., A. Grubler, V. Stojanovic, L. Riedl, and W. Pospischil. 2000. A GIS-based approach for modeling the spatial and temporal development of night-time lights. Pp. 389-394 in Angewandte Geographische Informationsverarbeitung XII, Beiträge zum AGIT-Symposium Salzburg, J. Strobl, T. Blaschke and G. Griesebner, eds. Heidelberg: Wichmann Verlag.

Pollack, J. B., D. Rind, A. Lacis, J. E. Hansen, M. Sato, and R. Ruedy. 1993. GCM simulations of volcanic aerosol forcing. Part 1: Climate changes induced by steady-state perturbations. Journal of Climate 6(9):1719-1742.

Poore, R. Z., H. J. Dowsett, and S. Verardo. 2003. Millennial- to century-scale variability in Gulf of Mexico Holocene climate records. Paleoceanography 180:2611-2613.

Porter, S. C., and G. H. Denton. 1967. Chronology of neoglaciation in the North American Cordillera. American Journal of Science 265:177-210.

Price, C., J. Penner, and M. Prather. 1997. NO_x from lightning: 1. Global distribution based on lightning physics. Journal of Geophysical Research 102(D5):5929-5941.

Quinn, P. K., and D. J. Coffman. 1998. Local closure during the First Aerosol Characterization Experiment (ACE 1): Aerosol mass concentration and scattering and backscattering coefficients. Journal of Geophysical Research 103:15575-15596.

Rabin, R. M., and D. W. Martin. 1995. Satellite observations of shallow cumulus coverage over the central United States: An exploration of land use impact on cloud cover. Journal of Geophysical Research—Atmospheres 101(D3):7149-7155.

Raddatz, R. L. 2003. Aridity and the potential physiological response of C_3 crops to doubled atmospheric CO_2: A simple demonstration of the sensitivity of the Canadian prairies. Boundary-Layer Meteorology 107:483-496.

Rahmstorf, S., D. Archer, D. S. Ebel, O. Eugster, J. Jouzel, D. Maraun, G. A. Schmidt, J. Severinghaus, A. J. Weaver, and J. Zachos. 2004. Cosmic rays, carbon dioxide, and climate. EOS, Transactions, American Geophysical Union 85:38, 41.

Ramachandran, S., V. Ramaswamy, G. L. Stenchikov, and A. Robock. 2000. Radiative impacts of the Mt. Pinatubo volcanic eruption: Lower stratospheric response. Journal of Geophysical Research 105:24409-24429.

Ramanathan, V., R. J. Cicerone, H. B. Singh, and J. T. Kiehl. 1985. Trace gas trends and their potential role in climate change. Journal of Geophysical Research 90:5547-5566.

Ramanathan, V., L. Callis, R. Cess, J. Hansen, I. Isaksen, W. Kuhn, A. Lacis, F. Luther, J. Mahlman, R. Reck, and M. Schlesinger. 1987. Climate-chemical interactions and effects of changing atmospheric trace gases. World Meteorological Organization Report 14. Reviews of Geophysics 25:1441-1482.

Ramanathan, V., E. F. Harrison, and B. R. Barkstrom. 1989. Climate and the Earth's radiation budget. Physics Today 42(5):22-33.

Ramanathan, V., B. Subasilar, G. J. Zhang, W. Conant, R. D. Cess, J. T. Kiehl, H. Grassl, and L. Shi. 1995. Warm pool heat budget and shortwave cloud forcing: A missing physics? Science 267:499-503.

Ramanathan, V., P. J. Crutzen, J. T. Kiehl, and D. Rosenfeld. 2001a. Aerosols, climate and the hydrological cycle. Science 294:2119-2124.

Ramanathan, V., P. J. Crutzen, J. Leleiveld, D. Althausen, J. Anderson, M. O. Andreae, W. Cantrell, G. Cass, C. E. Chung, A. D. Clarke, W. D. Collins, J. A. Coakley, F. Dulac, J. Heintzenberg, A. J. Heymsfield, B. Holben, J. Hudson, A. Jayaraman, J. T. Kiehl, T. N. Krishnamurti, D. Lubin, A. P. Mitra, G. MacFarquhar, T. Novakov, J. A. Ogren, I. A. Podgorny, K. Prather, J. M. Prospero, K. Priestley, P. K. Quinn, K. Rajeev, P. Rasch, S. Rupert, R. Sadourney, S. K. Satheesh, P. Sheridan, G. E. Shaw, and F. P. J. Valero. 2001b. The Indian Ocean Experiment: An integrated assessment of the climate forcing and effects of the great Indo-Asian haze. Journal of Geophysical Research 106(D22):28371-28399.

Ramankutty, N., and J. A. Foley. 1999. Estimating historical changes in global land cover: Croplands from 1700 to 1992. Global Biogeochemical Cycles 13:997-1027.

Ramaswamy, V., and C-T. Chen. 1997. Climate forcing-response relationships for greenhouse and shortwave radiative perturbations. Geophysical Research Letters 24:667-670.

Ramaswamy, V., and M. D. Schwarzkopf, 2002. Effects of ozone and well-mixed gases on annual-mean stratospheric temperature trends. Geophysical Research Letters, 29(22): 2064, DOI: 10.1029/2002GL015141.

Ramaswamy, V., M-L. Chanin, J. Angell, J. Barnett, D. Seidel, M. Gelman, P. Keckhut, Y. Koshlekov, K. Labitzke, J-J. Lin, A. O'Neill, J. Nash, W. Randel, R. Rood, K. Shine, M. Shiotani, and R. Swinbank 2001. Stratospheric temperature trends: Observations and model simulations. Reviews of Geophysics, 39:71-122.

Ramaswamy, V., S. Ramachandran, G. Stenchikov, and A. Robock. 2004. A model study of the effect of Pinatubo volcanic aerosols on stratospheric temperatures. In Frontiers in the Science of Climate Modeling, J. T. Kiehl and V. Ramanathan, eds. Cambridge, U.K.: Cambridge University Press.

Randel, D. L., T. H. Vonder Haar, M. A. Ringerud, G. L. Stephens, T. J. Greenwald, and C. L. Combs. 1996. A new global water vapor dataset. Bulletin of the American Meteorological Society 77:1233-1246.

Randles, C. A., L. M. Russell, and V. Ramaswamy. 2004. Hygroscopic and optical properties of organic sea-salt aerosol and consequences for climate forcing. Geophysical Research Letters 31(L16108), DOI: 10.1029/2004GL020628.

Raper, S. C. B., J. M. Gregory, and R. J. Stouffer. 2001. The role of climate sensitivity and ocean heat uptake on AOGCM transient temperature response. Journal of Climate 15:124-130.

Rasch, P. J., and J. E. Kristjánsson. 1998. A comparison of the CCM3 model climate using diagnosed and predicted consensate parameterizations. Journal of Climate 11:1587-1614.

Raymo, M. E. 1998. Glacial puzzles. Science 281:1467-1468.

Rayner, P. J., and D. M. O'Brien. 2001. The utility of remotely sensed CO_2 concentration data in surface source inversions. Geophysical Research Letters 28:175-178.

Redemann, J., R. P. Turco, K. N. Liou, P. B. Russell, R. W. Bergstrom, B. Schmid, J. M. Livingston, P. V. Hobbs, W. S. Hartley, R. Ismail, R. A. Ferrare, and E. V. Browell. 2000. Retrieving the vertical structure of the effective aerosol complex index of refraction from a combination of aerosol in situ and remote sensing measurements during TARFOX. Journal of Geophysical Research 105(D8):9949-9970.

Rial, J. A., R. A. Pielke Sr., M. Beniston, M. Claussen, J. Canadell, P. Cox, H. Held, N. De Noblet-Ducoudré, R. Prinn, J. F. Reynolds, and J. D. Salas. 2004. Nonlinearities, feedbacks and critical thresholds within the Earth's climate system. Climatic Change 65:11-38.

Rind, D., J. Lean, and R. Healy. 1999. Simulated time-dependent climate response to solar radiative forcing since 1600. Journal of Geophysical Research 104:1973-1990.

Robertson, A. D., J. T. Overpeck, E. Mosley-Thompson, G. A. Zielinski, J. L. Lean, D. Koch, J. E. Penner, I. Tegen, D. Rind, and R. Healy. 1998. Hypothesized climate forcing time series for the last 500 years (abstract). Supplement to EOS, Transactions American Geophysical Union 79:833-834.

Robock, A. 2000. Volcanic eruptions and climate. Reviews of Geophysics 38:191-219.

Robock, A. 2002. Volcanic eruptions. Pp. 738-744 in Encyclopedia of Global Environmental Change, Volume 1: The Earth System: Physical and chemical dimensions of global environmental change, M. C. MacCracken and J. S. Perry eds. Chichester: John Wiley and Sonts, Ltd.

Robock, A., and M. P. Free. 1995. Ice cores as an index of global volcanism from 1850 to the present. Journal of Geophysical Research 100:11549-11567.

Robock, A., and M. P. Free. 1996. The volcanic record in ice cores for the past 2000 years. Pp. 533-546 in Climatic Variations and Forcing Mechanisms of the Last 2000 Years, P. D. Jones, R. S. Bradley, and J. Jouzel, eds. Berlin: Springer Verlag.

Roderick, M. L., and G. D. Farquhar. 2002. The cause of decreased pan evaporation over the past 50 years. Science 298:1410-1411

Roehrl, R. A., and K. Riahi, 2000. Greenhouse gas emissions mitigation and the role of technology dynamics and path dependency—A cost assessment. Technological Forecasting and Social Change 63(2-3).

Rosenfeld, D. 1999. Dense smoke turns off normal tropical rainfall. Geophysical Research Letters 26:3105.

Rosenfeld, D. 2000. Suppression of rain and snow by urban and industrial air pollution. Science 287:1793-1796.

Rosenfeld, D., and G. Feingold. 2003. Explanation of discrepancies among satellite observations of the aerosol indirect effects. Geophysical Research Letters 30(14):1776, DOI: 10.1029/2003GL017684.

Rosenfeld, D., and W. L. Woodley. 2000. Convective clouds with sustained highly supercooled liquid water down to −37.5°C. Nature 405:440-442.

Rosenlof, K. H., S. J. Oltmans, D. Kley, J. M. Russell III, E.-W. Chiou, W. P. Chu, D. G. Johnson, K. K. Kelly, H. A. Michelsen, G. E. Nedoluha, E. E. Remsberg, G. C. Toon, and M. P. McCormick. 2001. Stratospheric water vapor increases of the past half-century. Geophysical Research Letters 28:1195-1198.

Rotstayn, L. D., and Y. G. Liu. 2003. Sensitivity of the first indirect aerosol effect to an increase of cloud droplet spectral dispersion with droplet number concentration. Journal of Climate 16(21):3476-3481.

Royer, D. L., R. A. Berner, I. P. Montañez, N. J. Tabor, and D. J. Beerling. 2004. CO_2 as a primary driver of Phanerozoic climate. GSA Today 14(3):4-10.

Ruddiman, W. F. 2001. Earth's Climate, Past and Future. New York: W.H. Freeman and Company. 465 pp.

Ruddiman, W. F. 2003. The anthropogenic greenhouse era began thousands of years ago. Climatic Change 61:261-293.

Running, S. W., R. Nemani, F. Heinsch, M. Zhao, M. Reeves, and H. Hashimoto. 2004. A continuous satellite-derived measure of global terrestrial primary productivity: Future science and applications. Bioscience 54(6):547-560.

Russell, L. M. 2003. Aerosol organic-mass-to-organic-carbon ratio measurements. Environmental Science and Technology 37(13):2982-2997.

Russell, L. M., and J. H. Seinfeld. 1998. Size- and composition-resolved externally mixed aerosol model. Aerosol Science and Technology 28(5):403-416.

Russell, L. M., J. H. Seinfeld, R. C. Flagan, R. J. Ferek, D. A. Hegg, P. V. Hobbs, W. Wobrock, A. I. Flossmann, C. D. O'Dowd, K. E. Nielsen, and P. A. Durkee. 1999. Aerosol dynamics in ship tracks. Journal of Geophysical Research—Atmospheres 104(D24):31077-31095.

Russell, L. M., K. J. Noone, R. J. Ferek, R. A. Pockalny, R. C. Flagan, and J. H. Seinfeld. 2000. Combustion organic aerosol as cloud condensation nuclei in ship tracks. Journal of the Atmospheric Sciences 57(16):2591-2606.

Russell, L. M., S. F. Maria, and S. Myneni. 2002. Mapping organic coatings on atmospheric particles. Geophysical Research Letters 29, DOI: 10.1029/2002GL014874.

Russell, P. B., J. M. Livingston, P. Hignett, S. Kinne, J. Wong, A. Chien, R. Bergstrom, P. Durkee, and P. V. Hobbs. 1999. Aerosol-induced radiative flux changes off the United States mid-Atlantic coast: Comparison of values calculated from sunphotometer and in situ data with those measured by airborne pyranometer. Journal of Geophysical Research—Atmospheres 104(D2):2289-2307.

Ruzmaikin, A. 1999. Can El Niño amplify the solar forcing of climate? Geophysical Research Letters 26:2255-2258.

Ruzmaikin, A., and J. Feynman. 2002. Solar influence on a major mode of atmospheric variability. Geophysical Research Letters 107(D14).

Rye, R., P. H. Kuo, and H. D. Holland. 1995. Atmospheric carbon dioxide concentrations before 2.2 billion years ago. Nature 378:603-605.

Sagan, C., and C. Chyba. 1997. The early faint Sun paradox: Organic shielding of ultraviolet-labile greenhouse gases. Science 276:1217-1221.

Sagan, C., and G. Mullen. 1972. Earth and Mars—Evolution of Atmospheres and Surface Temperatures. Science 177:52.

Salby, M., and P. Callaghan. 2000. Connection between the solar cycle and the QBO: The missing link. Journal of Climate 13:2652-2662.

Sanadze, G. A. 1969. Light-dependent excretion of molecular isoprene. Progress in Photosynthesis Research 2:701-706.

Santer, B. D., K. E. Taylor, J. E. Penner, T. M. L. Wigley, U. Cubasch, and P. D. Jones. 1995. Towards the detection and attribution of an anthropogenic effect on climate. Climate Dynamics 12:77-100.

Santer, B. D., K. E. Taylor, T. M. L. Wigley, P. D. Jones, D. J. Karoly, J. F. B. Mitchell, A. H. Oort, J. E. Penner, V. Ramaswamy, M. D. Schwarzkopf, R. S. Stouffer, and S. F. B. Tett. 1996. A search for human influences on the thermal structure in the atmosphere. Nature 382:39-46.

Santer, B. D., T. M. L. Wigley, D. J. Gaffen, L. Bengtsson, C. Doutriaux, J. S. Boyle, M. Esch, J. J. Hnilo, P. D. Jones, G. A. Meehl, E. Roeckner, K. E. Taylor, and M. F. Wehner. 2000. Interpreting differential temperature trends at the surface and in the lower troposphere. Science 287:1227-1232.

Santer, B. D., T. M. L. Wigley, G. A. Meehl, M. F. Wehner, C. Mears, M. Schabel, F. J. Wentz, C. Ammann, J. Arblaster, T. Bettge, W. M. Washington, K. E. Taylor, J. S. Boyle, W. Brüggemann, and C. Doutriaux. 2003a. Influence of satellite data uncertainties on the detection of externally forced climate change. Science 300:1280-1284.

Santer, B. D., M. F. Wehner, T. M. L. Wigley, R. Sausen, G. A. Meehl, K. E. Taylor, C. Ammann, J. Arblaster, W. M. Washington, J. S. Boyle, and W. Brüggemann. 2003b. Contributions of anthropogenic and natural forcing to recent tropopause height changes. Science 301:479-483.

Sarewitz, D., R. A. Pielke Jr., and R. Byerly Jr., eds. 2000. Prediction: Science, decision making and the future of nature. Washington, D.C.: Island Press.

Sassen K., P. J. DeMott, J. M. Prospero, and M. R. Poellot. 2003. Saharan dust storms and indirect aerosol effects on clouds: CRYSTAL-FACE results. Geophysical Research Letters 30(12):1633, DOI: 10.1029/2003GL017371.

Satheesh, S. K., and V. Ramanathan. 2000. Large differences in tropical aerosol forcing at the top of the atmosphere and Earth's surface. Nature 405(6782):60-63.

Schaaf, C. B., A. H. Strahler, F. Gao, W. Lucht, X. Li, X. Zhang, Y. Jin, E. Tsvetsinskaya, J.-P. Muller, P. Lewis, M. Barnsley, G. Roberts, C. Doll, S. Liang, J. L. Privette, and D. Roy. 2002. Global Albedo, BRDF and Nadir BRDF-adjusted reflectance products from MODIS. Proceedings of the International Geoscience and Remote Sensing Symposium (IGARSS'02), Toronto, Canada, June 24-28.

Schauer, J. J., M. J. Kleeman, G. R. Cass, and B. R. T. Simoneit. 1999. Measurement of emissions from air pollution sources. 2. C_1-C_{29} organic compounds from medium duty diesel trucks. Environmental Science and Technology 33:1578-1587.

Schimel, D. S., J. I. House, K. A. Hibbard, P. Bousquet, P. Ciais, P. Peylin, B. H. Braswell, M. J. Apps, D. Baker, A. Bondeau, J. Canadell, G. Churkina, W. Cramer, A. S. Denning, C. B. Field, P. Friedlingstein, C. Goodale, M. Heimann, R. A. Houghton, J. M. Melillo, B. Moore III, D. Murdiyarso, I. Noble, S. W. Pacala, I. C. Prentice, M. R. Raupach, P. J. Rayner, R. J. Scholes, W. L. Steffen, and C. Wirth. 2001. Recent patterns and mechanisms of carbon exchange by terrestrial ecosystems. Nature 414:169-172.

Schipper, L. 1996. Lifestyles and the environment: The case for energy. Daedalus 125:113.

Schmidt, G. A., and D. T. Shindell. 2003. Atmospheric composition, radiative forcing, and climate change as a consequence of a massive methane release from gas hydrates. Paleoceanography 18(1):1004, DOI: 10.1029/2002PA000757.

Schneider, S. H., and R. E. Dickinson. 1974. Climate modeling. Review of Geophysics and Space Physics 12:447-493.

Schneider, S. H., and C. Mass. 1975. Volcanic dust, sunspots and temperature trends. Science 190:741-746.

Schwartz, S. E. 1993. Does fossil fuel combustion lead to global warming? Energy International Journal 18:1229-1248.

Schwartz, S. E., Harshvardhan, and C. M. Benkovitz. 2002. Influence of anthropogenic aerosol on cloud optical depth and albedo shown by satellite measurements and chemical transport modeling. Proceedings of the National Academy of Sciences 99(4):1784-1789.

Schwarzkopf, M. D., and V. Ramaswamy. 2002. Effects of changes in well-mixed gases and ozone on stratospheric seasonal temperatures. Geophysical Research Letters, 29(24):2184, DOI: 10.1029/2002GL015759.

Seinfeld, J. H., and S. N. Pandis. 1998. Atmospheric Chemistry and Physics: From Air Pollution to Climate Change. New York: John Wiley & Sons.

Self, S., T. Thordarson, and L. Keszthelyi. 1997. Emplacement of continental basalt lava flows. Pp. 381-410 in Large Igneous Provinces: Continental, Oceanic, and Planetary Flood Volcanism, J. J. Mahoney and M. F. Coffin, eds. Geophysical Monograph 100. Washington, D.C.: American Geophysical Union.

Sellers, P. J., R. E. Dickinson, D. A. Randall, A. K. Betts, F. G. Hall, J. A. Berry, G. J. Collatz, A. S. Denning, H. A. Mooney, C. A. Nobre, N. Sato, C. B. Field, and A. Henderson-Sellers. 1997. Modeling the exchanges of energy, water, and carbon between continents and the atmosphere. Science 275(5299):502-509.

Sexton, D. M. H., H. Grubb, K. P. Shine, and C. K. Folland. 2003. Design and analysis of climate model experiments for the efficient estimation of anthropogenic signals. Journal of Climate 16:1320-1336.

Shaviv, N. J. 2003. Toward a solution of the early faint Sun paradox: A lower cosmic ray flux from a stronger solar wind. Journal of Geophysical Research 108(A12), DOI: 10.1029/2003JA009997.

Shaviv, N. J., and J. Veizer. 2003. Celestial driver of Phanerozoic climate? GSA Today 13(7):4-10.

Shell, K. M., R. Frouin, S. Nakamoto, and R. C. J. Somerville. 2003. Atmospheric response to solar radiation absorbed by phytoplankton. Journal of Geophysical Research 108(D15):4445.

Sherwood, S. 2002. A microphysical connection among biomass burning, cumulus clouds, and stratospheric moisture. Science 295:1272-1275.

Shindell, D. T., and G. Faluvegi. 2002. An exploration of ozone changes and their radiative forcing prior to the chlorofluorocarbon era. Atmospheric Chemistry and Physics 2:363-374.

Shindell D. T., D. Rind, N. Balachandran, J. Lean, and P. Lonergan. 1999. Solar cycle variability, ozone, and climate. Science 284(5412): 305-308.

Shindell, D. T., G. A. Schmidt, R. L. Miller, and D. Rind. 2001a. Northern hemisphere winter climate response to greenhouse gas, ozone, solar, and volcanic forcing. Journal of Geophysical Research 106(D7):7193-7210.

Shindell, D. T., G. A. Schmidt, M. E. Mann, D. Rind, and A. Waple. 2001b. Solar forcing of regional climate change during the Maunder Minimum. Science 294:2149-2152.

Shindell, D. T., G. A. Schmidt, R. L. Miller, and M. E. Mann. 2003. Volcanic and solar forcing of climate during the preindustrial era. Journal of Climate 16:4094-4107.

Shindell, D. T., G. A. Schmidt, M. E. Mann, and G. Faluvegi. 2004. Dynamic winter climate response to large tropical volcanic eruptions since 1600. Journal of Geophysical Research 109D05104, DOI:10.1029/2003JD004151.

Shine, K. P., J. Cook, E. J. Highwood, and M. M. Joshi. 2003. An alternative to radiative forcing for estimating the relative importance of climate change mechanisms. Geophysical Research Letters 30(20):2047, DOI: 10.1029/ 2003GL018141.

Shukla, J., and Y. Mintz. 1982. Influence of land-surface evapotranspiration on the Earth's climate. Science 215:1498-1501.

Smith, L. C., G. M. MacDonald, A. A. Velichko, D. W. Beilman, O. K. Borisova, K. E. Frey, R. V. Kremenetski, and Y. Sheng. 2004. Siberian peatlands a net carbon sink and global methane source since the early Holocene. Science 303:353-356.

Smith, S. J., and T. M. L. Wigley. 2000a. Global warming potentials. 1: Climatic implications of emissions reductions. Climatic Change 44:445-457.

Smith, S. J., and T. M. L. Wigley. 2000b. Global warming potentials. 2: Accuracy. Climatic Change 44:459-469.

Snyder, P. K., J. A. Foley, M. H. Hitchman, and C. Delire. 2004. Analyzing the effects of complete tropical forest removal on the regional climate using a detailed three-dimensional energy budget: An application to Africa. Journal of Geophysical Research-Atmospheres 109(D21):D21102, DOI: 10.1029/2003JD004462

Soden, B. J., and V. Ramaswamy. 1998. Variations in atmosphere-ocean solar absorption under clear skies: A comparison of observations and models. Geophysical Research Letters 25(12):2149-2152.

Soden, B. J., R. T. Wetherald, G. Stenchikov, and A. Robock. 2002. Global cooling after the eruption of Mount Pinatubo: A test of climate feedback by water vapor. Science 296(5568):727-730.

Sohl, T. L, A. L. Gallant, and T. R. Loveland. 2004. The characteristics and interpretability of land surface change and implications on project design. Photogrammetric Engineering and Remote Sensing 70:439-450.

Sollazzo, M. J., L. M. Russell, D. Percival, S. Osborne, R. Wood, and D. W. Johnson. 2000. Entrainment rates during ACE 2 Lagrangian experiments calculated from aircraft measurements. Tellus 52B:335-347.

Soubottina, T. 2004. Beyond Economic Growth: An Introduction to Sustainable Development, 2nd ed. Washington, D.C.: The World Bank.

Spessa, A., S. P. Harrison, I. C. Prentice, W. Cramer, N. Mahowald. 2003. Confronting a burning question: The role of eire on Earth. Eos Trans. American Geophysical Union. 84(3):23, 10.1029/2003EO030005.

Srinivisan, K., and V. Shastri. 2004. A Set of Population Projections of India and the Larger States Based on 2001 Census Results. Population Foundation of India. Available on-line at *http://planningcommission.nic.in/reports/genrep/bkpap2020/25_bg2020.doc*. Accessed June 21, 2004.

Stajner, I., L. P. Riishøjgaard, and R. B. Rood. 2001. The GEOS ozone data assimilation system: Specification of error statistics. Quarterly Journal of the Royal Meteorological Society 127:1069-1094.

Stanhill, G., and S. Cohen. 2001. Global dimming: A review of the evidence for a widespread and significant reduction in global radiation with discussion of its probable causes and possible agricultural consequences. Agriculture Forest Meteorolology 107:255-278.

Staudt, A. C., D. J. Jacob, F. Ravetta, J. A. Logan, D. Bachiochi, T. N. Krishnamurti, S. T Sandholm, B. A. Ridley, H. B. Singh, and R. W. Talbot. 2003. Sources and chemistry of nitrogen oxides over the tropical Pacific. Journal of Geophysical Research 108:8239, DOI: 10.1029/2002JD002139.

Steffen, W, A. Sanderson, P. D. Tyson, J. Jäger, P. A. Matson, B. Moore III, F. Oldfield, K. Richardson, H. J. Schellnhuber, B. L. Turner II, and R. J. Wasson. 2004. Global Change and the Earth System: A Planet Under Pressure. Executive Summary. Stockholm: International Geosphere-Biosphere Program.

Steiger, S. M., and R. E. Orville. 2003. Cloud-to-ground lightning enhancement over southern Louisiana. Geophysical Research Letters 29, DOI: 10.1029/2003GL017923.

Stenchikov, G. L., I. Kirchner, A. Robock, H.-F. Graf, J.-C. Antuña, R. G. Grainger, A. Lambert, and L. Thomason. 1998. Radiative forcing from the 1991 Mount Pinatubo volcanic eruption. Journal of Geophysical Research 103:13837-13857.

Stenchikov, G. L., A. Robock, V. Ramaswamy, M. D. Schwarzkopf, K. Hamilton, and S. Ramachandran. 2002. Arctic Oscillation response to the 1991 Mount Pinatubo eruption: Effects of volcanic aerosols and ozone depletion. Journal of Geophysical Research 107, DOI: 10.1029/2002JD002090.

Steyaert, L. T., F. G. Hall, and T. R. Loveland. 1997. Land cover mapping, fire disturbance-regeneration, and multiresolution land cover scaling studies in the BOREAS forest ecosystem with multiresolution 1-km AVHRR. Journal of Geophysical Research 102:29581-29598.

Stott, P. A., S. F. B. Tett, G. S. Jones, M. R. Allen, J. F. B. Mitchell, and G. J. Jenkins. 2000. External control of 20th century temperature by natural and anthropogenic forcings. Science 290:2133-2137.

Stott, P. A., S. F. B. Tett, G. S. Jones, M. R. Allen, W. J. Ingram, and J. F. B. Mitchell. 2001. Attribution of twentieth century temperature change to natural and anthropogenic causes. Climate Dynamics 17:1-21.

Strahler, A., D. Muchoney, J. Borak, and M. Friedl. 1999. MODIS land cover product algorithm theoretical basis document (ATBD) version 5.0. Available on-line at *http://modis-land.gsfc.nasa.gov/products/products.asp?ProdFamID=10*. Accessed December 9, 2004.

Ström, J., and S. Ohlsson. 1998. In situ measurements of enhanced crystal number densities in cirrus clouds caused by aircraft exhaust. Journal of Geophysical Research 103(D10):11355-11362, DOI: 10.1029/98JD00807.

Stuber, N., M. Ponater, and R. Sausen. 2001. Is the climate sensitivity to ozone perturbations enhanced by stratospheric water vapor feedback? Geophysical Research Letters 28(15), DOI: 10.1029/2001GL013000. issn: 0094-8276.

Stuiver, M. 1965. Carbon-14 content of 18th- and 19th-century wood variations correlated with sunspot activity. Science 149:533-34.

Suntharalingam, P., D. J. Jacob, P. I. Palmer, J. A. Logan, R. M. Yantosca, Y. Xiao, M. J. Evans, D. G. Streets, S. L. Vay, and G. W. Sachse. 2004. Improved quantification of Chinese carbon fluxes using CO_2/CO correlations in Asian outflow. Journal of Geophysical Research 109(18):D18S18, DOI: 10.1029/2003JD004362.

Svensmark, H., and E. Friis-Christensen. 1997. Variation of cosmic ray flux and global cloud coverage—A missing link in solar-climate relationships. Journal of Atmospheric and Solar-Terrestrial Physics 59:1225-1232.

Tan, W. W., M. A. Geller, S. Pawson, and A. da Silva. 2004. A case study of excessive subtropical transport in the stratosphere of a data assimilation system. Journal of Geophysical Research 109:D11102, DOI:10.1029/2003JD004057.

Taylor, J. P., and A. McHaffie. 1994. Measurements of cloud susceptibility. Journal of the Atmospheric Sciences 51:1298-1306.

Teller, J., and D. Leverington. 2004. Glacial Lake Agassiz: A 5000-year history of change and its relationship to the ^{18}O record of Greenland. Geological Society of America Bulletin 116:729-742.

Tett, S. F. B., J. F. B. Mitchell, D. E. Parker, and M. R. Allen. 1996. Human influence on the atmospheric vertical temperature structure: Direction and observations. Science 274:1170-1173.

Tett, S. F. B., P. A. Scott, M. R. Allen, W. J. Ingram, and J. F. B. Mitchell. 1999. Causes of twentieth-century temperature change near the Earth's surface. Nature 399:569-572.

Thompson, L. G., E. Mosley-Thompson, M. E. Davis, K. A. Henderson, H. H. Brecher, V. S. Zagorodnov, T. A. Mashiotta, P.-N. Lin, V. N. Mikhalenko, D. R. Hardy, and J. Beer. 2002. Kilimanjaro ice core records: Evidence of Holocene climate change in tropical Africa. Science 298:589-593.

Thordarson, T., S. Self, D. J. Miller, G. Larsen, and E. G. Vilmundardottir. 2003. Sulphur release from lava eruptions in the Veidivotn, Grimsvotn and Katla volcanic systems, Iceland. Pp. 103-121 in Volcanic Degassing, C. Oppenheimer, D. M. Pyle, and J. Barclay, eds. Special Publications 213. London: Geological Society of London.

Tie, X., R. Zhang, G. Brasseur, and W. Lei. 2002. Global NO_x production by lightning. Journal of Atmospheric Chemistry 43:61-74.

Toth, F., G.-Y. Cao, and E. Hizsnyik. 2003. Regional Population Projections for China: Interim Report IR-03-042. Laxenburg, Austria: International Institute for Applied Systems Analysis (IIASA). Available on-line at *http://www.iiasa.ac.at/Publications/Documents/IR-03-042.pdf*. Accessed June 21, 2004.

Townshend, J. R. G., and C. O. Justice. 2002. Towards operational monitoring of terrestrial systems by moderate resolution remote sensing. Remote Sensing of Environment 83:351-359.

Travis, D. J., A. M. Carleton, and R. Lauritsen. 2002. An unintended contrail experiment. Bulletin of the American Meteorological Society 83(4):500.

Trenberth, K. E. 2004. Rural land-use change and climate. Nature 427:213.

Tripati, A. K., and H. Elderfield. 2004. Abrupt hydrographic changes in the equatorial Pacific and subtropical Atlantic from foraminiferal Mg/Ca indicate greenhouse origin for the thermal maximum at the Paleocene-Eocene boundary. Geochemistry Geophysics Geosystems 5:1-ll.

Tsonis, A. A. 2001. The impact of nonlinear dynamics in the atmospheric sciences. International Journal of Bifurcation and Chaos 11:881-902.

Twomey, S. A. 1959. The nuclei of natural cloud formation. Part II: The supersaturation in natural clouds and the variation of cloud droplet concentrations. Geofisica Pura e Applicata 43:227-242.

Udelhofen, P. M., and R. D. Cess. 2001. Cloud cover variations over the United States: An influence of cosmic rays or solar variability? Geophysical Research Letters 28:2617-2620.

UN (United Nations). 2002. United Nations World Population Prospects: The 2002 Revision. New York: United Nations.

USCCSP (U.S. Climate Change Science Program). 2003. Strategic Plan for the U.S. Climate Change Science Program. Washington, D.C.

U.S. Census Bureau. 2002. State Population Projections. Available on-line at *http://www. census.gov/population/www/projections/stproj.html*. Accessed June 21, 2004.

Van Aardenne, J. A., F. J. Dentener, J. G. J. Olivier, C. G. M. Klein Goldewijk, and J. Lelieveld. 2001. A $1° \times 1°$ resolution data set of historical anthropogenic trace gas emissions for the period 1890-1990. Global Biogeochemical Cycles 15(4):909-928.

Van den Hurk, B. J. J. M., P. Viterbo, and S. O. Los. 2003. Impact of leaf area index seasonality on the annual land surface evaporation in a global circulation model. Journal of Geophysical Research—Atmospheres 108(D6):4191.

Vinnikov, K. Y. and N. C. Grody. 2003: Global warming trend of mean tropospheric temperature observed by satellites. Science 302:269-272.

Vose, R. S., T. R. Karl, D. R. Easterling, C. N. Williams, and M. J. Menne. 2004. Impact of land-use change on climate. Nature 427:213-214.

Wang J., S. A. Christopher, J. S. Reid, H. Maring, D. Savoie, B. N. Holben, J. M. Livingston, P. B. Russell, and S.-K. Yang. 2003. GOES 8 retrieval of dust aerosol optical thickness over the Atlantic Ocean during PRIDE. Journal of Geophysical Research 108(D19):8595, DOI: 10.1029/2002JD002494.

Wang, J. S., J. A. Logan, M. B. McElroy, B. N. Duncan, I. A. Megretskaia, and R. M. Yantosca. 2004. A 3-D model analysis of the slowdown and interannual variability in the methane growth rate from 1988 to 1997. Global Biogeochemical Cycles 19, DOI: 10.1029/2003GB002180.

Wang, Y., and D. J. Jacob. 1998. Anthropogenic forcing on tropospheric ozone and OH since preindustrial times. Journal of Geophysical Research 103:31123-31135.

Wang, Y., D. J. Jacob, and J. A. Logan. 1998. Global simulation of tropospheric O_3-NO_x-hydrocarbon chemistry. 1. Model formulation. Journal of Geophysical Research 103(D9):10713-10726.

Waple, A. M., M. E. Mann, and R. S. Bradley. 2002. Long-term patterns of solar irradiance forcing in model experiments and proxy based surface temperature reconstructions. Climate Dynamics 18:563-578.

Warren, D. R., and J. H. Seinfeld. 1985. Simulation of aerosol size distribution evolution in systems with simultaneous nucleation, condensation, and coagulation. Aerosol Science and Technology 4(1):31-43.

Weaver, C. P., and R. Avissar. 2001. Atmospheric disturbances caused by human modification of the landscape. Bulletin of the American Meteorological Society 82:269-281.

Webber, W. R., and P. R. Higbie. 2003. Production of cosmogenic Be nuclei in the Earth's atmosphere by cosmic rays: Its dependence on solar modulation and the interstellar cosmic ray spectrum. Journal of Geophysical Research 108, DOI: 10.1029/2003JA009863.

Whitworth III, T., S. B. Rutz, R. D. Pillsbury, M. I. Moore, B. A. Warren, and W. D. Nowlin Jr. 1999. On the deep western-boundary current in the Southwest Pacific Basin. Progress in Oceanography 43(1):1-54.

Wigley, T. M. L. 1991. Could reducing fossil-fuel emissions cause global warming? Nature 349:503-506.

Wigley, T. M. L. 1998. The Kyoto Protocol: CO_2, CH_4 and climate implications. Geophysical Research Letters 25(13):2285-2288.

Wigley, T. M. L., and S. C. B. Raper. 2001. Interpretation of high projections for global-mean warming. Science 293:451-454, DOI: 10.1126/science.1061604.

Wigley, T. M. L., B. D. Santer, and K. E. Taylor. 2000. Correlation approaches to detection. Geophysical Research Letters 27(18):2973-2976.

Willis J. K., D. Roemmich, and B. Cornuelle, 2003. Combining altimetric height with broadscale profile data to estimate steric height, heat storage, subsurface temperature, and sea-surface temperature variability. Journal of Geophysical Research 108(C9):3292, DOI: 10.1029/2002JC001755.

Willis, J. K., D. Roemmich, and B. Cornuelle. 2004. Interannual variability in upper-ocean heat content, temperature, and thermosteric expansion on global scales. Journal of Geophysical Research 109(C12036), DOI:10.1029/2003JC002260.

Willson, R. C., and A. Mordvinov. 2003. Secular total irradiance trends during solar cycles 21-23. Geophysical Research Letters 30:1029-1032.

WMO (World Meteorological Organization). 1985. Atmospheric Ozone: Assessment of Our Understanding of the Processes Controlling its Present Distribution and Change. 3 vol. WMO Report No. 16. Geneva.

WMO. 2003. Scientific Assessment of Ozone Depletion: 2002, Global Ozone Research and Monitoring Project—Report No. 47, 498 pp., Geneva.

Wofsy, S. C., and R. C. Harriss. 2002. The North American Carbon Program (NACP). Report of the NACP Committee of the U.S. Interagency Carbon Cycle Science Program. Washington, D.C.: U.S. Global Change Research Program.

Workshop on GHG Stabilization Scenarios. 2004. Tsukuba, Japan, January 22-23.

World Bank. 2004. World Development Indicators 2004. Washington, D.C.: World Bank.

World Energy Outlook. 2004. Available on-line at *http://www.worldenergyoutlook.org/*. Accessed September 2004.

Wu, Z.-X., and R. E. Newell. 1998. Influence of sea surface temperature on air temperature in the tropic. Climate Dynamics 14:275-290.

Wuchterl, G., and R. S. Klessen. 2001. The first million years of the Sun: A calculation of the formation and early evolution of a solar mass star. Astrophysical Journal 560:L185-L188.

Wunsch, C. 2004. Quantitative estimate of the Milankovitch-forced contribution to observed Quaternary climate change. Quaternary Science Reviews 23:1001-1012.

Xiao, Y., D. J. Jacob, J. S. Wang, J. A. Logan, P. I. Palmer, P. Suntharalingam, R. M. Yantosca, G. W. Sachse, D. R. Blake, and D. G. Streets. 2004. Constraints on Asian and European sources of methane from CH_4-C_2H_6-CO correlations in Asian outflow. Journal of Geophysical Research 109:D15S16, DOI: 10.1029/2003JD004475.

Yamamoto, G., and M. Tanaka. 1972. Increase of global albedo due to air pollution. Journal of Atmospheric Science 29:1405-1412.

Yoder, J. A., J. K. Moore, and R. N. Swift. 2001. Putting together the big picture: Remote-sensing observations of ocean color. Oceanography 14(4):33-40.

Zielinski, G. A. 2000. Use of paleo-records in determining variability within the volcanism-climate system. Quaternary Science Reviews 19:417-438.

A

Biographical Sketches of Committee Members and Staff

COMMITTEE MEMBERS

Dr. Daniel J. Jacob is the Gordon McKay Professor of Atmospheric Chemistry and Environmental Engineering at Harvard University. His research focuses on understanding the composition of the atmosphere, its perturbation by human activity, and the implications for human welfare and climate. Dr. Jacob serves on the National Aeronautics and Space Administration (NASA) Earth Systems Science and Applications Advisory Committee (ESSAAC) and has been lead or co-lead scientist on several NASA aircraft missions. He is also the lead scientist for the GEOS-CHEM chemical transport model used by a large number of research groups in North America and Europe. He is the recipient of the NASA Distinguished Public Service Medal (2003) and the American Geophysical Union (AGU) James B. Macelwane Medal (1994). Dr. Jacob earned his Ph.D. in environmental engineering at the California Institute of Technology. He has previously served on the National Research Council (NRC) Committee on Earth Sciences and the Committee for the Study on Transportation and a Sustainable Environment.

Dr. Roni Avissar is the W. H. Gardner Professor and chair of the Department of Civil and Environmental Engineering at Duke University. He received his Ph.D. in 1987 from the Hebrew University, where he studied soil and water sciences and atmospheric sciences. His research focuses on the study of land-atmosphere interactions from micro to global scales, including the development and use of a variety of atmospheric, land, and oceanic models. Before joining Duke in 2001, he was at Rutgers University, where he started his academic career in 1989. Dr. Avissar served as editor of the *Journal of Geophysical Research—Climate and Physics of the Atmosphere.* Dr. Avissar has served on various national and international panels and committees including the NRC's Committee on Hydrologic Science. He currently serves as the project scientist for the hydrometeorology component of the Large-scale Biosphere Atmosphere (LBA) Experiment in the Amazon and is the chairman of the U.S. Global Change Research Program (USGCRP) Global Water Cycle Science Steering Group.

Dr. Gerard C. Bond is a Doherty senior scholar at Lamont-Doherty Earth Observatory of Columbia University, New York. Since 1980 he has been working on the history of the Earth's climate, mainly from the present through the previous interglaciation. His research interests have included the origin of Heinrich events, Dansgaard/Oeschger cycles, and the persistent 1500-year climate cycle. He is currently working on abrupt climate change within interglacial climates—particularly our present interglacial, or Holocene—and on how the Sun impacts the Earth's climate system. Dr. Bond is a fellow of the Geological Society of America and the recipient of the 2003 Maurice Ewing Medal. Dr. Bond received his Ph.D. in geology with minors in marine geology and geochemistry from the University of Wisconsin, Madison.

Dr. Stuart Gaffin is an associate research scientist at Columbia University's Center for Climate Systems Research. Previously, he was a senior scientist for the Global and Regional Atmospheric Program at Environmental Defense (formerly EDF). Dr. Gaffin's research focuses on emissions scenarios for greenhouse gases over the next century. He served as a lead author for the Intergovernmental Panel on Climate Change (IPCC) *Special Report on Emissions Scenarios.* He was a consulting scientist with the World Commission on Dams and focused on quantifying greenhouse gas emissions from flooded vegetation in dam reservoirs in the tropics of Brazil. Currently, Dr. Gaffin is specializing in the nexus between climate change, population and development, and environmental sustainability. Dr. Gaffin received his Ph.D. in climatology and geophysics from New York University's Earth Systems Group.

Dr. Jeffrey T. Kiehl is a senior scientist at the National Center for Atmospheric Research's (NCAR's) Climate Change and Research Section. This section applies the Community Climate System Model (CCSM) to past, present, and future climate change. Dr. Kiehl has carried out research on the effects of ozone depletion on Earth's climate, the role of clouds in the climate system, and the role of aerosol particles in the climate system. For the past two years, Dr. Kiehl has been chairman of the Scientific Steering Committee for the CCSM and led the development of the CCSM modeling effort. He also led a CCSM effort to simulate the climate of the twentieth century, including the effects of greenhouse gases. He was a contributing author to the chapters on aerosols and radiative forcings in the IPCC Third Assessment Report. He has served on the Climate Research Committee of the National Research Council, as editor for the *Journal of Geophysical Research*, and as a member of the Board of Reviewing Editors for *Science* magazine. He has also served on the Science Steering Committee for the U.S. Climate Variability (CLIVAR) board of the National Research Council. Dr. Kiehl received his Ph.D. in atmospheric science from the State University of New York at Albany.

Dr. Judith L. Lean is a research physicist at the Naval Research Laboratory. She received her Ph.D. in atmospheric physics from the University of Adelaide, Australia. She specializes in the study of the variability of solar radiation and its impact on Earth's climate and space weather. The focus of her current research is the mechanisms, models, and measurements of variation in the Sun's radiative output. Dr. Lean served as the chair of a group of scientists who assisted the National Research Council Board on Global Change to prepare the 1994 report *Solar Influences on Global Change*. She is a member of the National Academy of Sciences and has also served on the NRC's Board on Atmospheric Sciences and Climate, Committee for a Review of Scientific Aspects of the NASA Triana Mission, and the Task Group on Ground-Based Solar Research.

Dr. Ulrike Lohmann is a full professor and leads the Atmospheric Physics group at ETH Zurich, Switzerland. Until recently Dr. Lohmann was an associate Professor, Canada Research Chair and Coordinator of the Atmospheric Science Program in the Department of Physics and Atmospheric Science at Dalhousie University. Her research activities concentrate on the role of clouds and aerosols in the climate system. Dr. Lohmann is a member of the scientific advisory committee for SOLAS (Surface Ocean Lower Atmosphere Study) Canada, a member of the scientific steering committee of the International Global Atmospheric Chemistry (IGAC) project and a member of the International Commission of Clouds and Precipitation (ICCP). She was a contributing author for multiple chapters of the Intergovernmental Panel on Climate Change Third Assessment Report. Dr. Lohmann received her Ph.D. in Meteorology from the Max Planck Institute for Meteorology/Hamburg University, Germany.

Dr. Michael E. Mann is an assistant professor in the Department of Environmental Sciences at the University of Virginia. Dr. Mann earned his

Ph.D. from the Department of Geology and Geophysics at Yale University. Dr. Mann's research focuses on the application of statistical techniques to understanding climate variability and climate change from both empirical and climate model-based perspectives. A specific area of current research is paleoclimate data synthesis and statistically based climate pattern reconstruction during past centuries using climate "proxy" data networks. A primary focus of this research is deducing empirically the long-term behavior of the climate system and its relationship with possible external (including anthropogenic) "forcings" of climate. Dr. Mann was a lead author on the "Observed Climate Variability and Change" chapter of the IPCC Third Assessment Report. He was an invited participant in the 2002 National Research Council workshop Estimating Climate Sensitivity and is the current organizing committee chair for the National Academy of Sciences Frontiers of Science symposium. He currently serves as an editor of the *Journal of Climate*, and is a participant in numerous other scientific committees and working groups.

Dr. Roger A. Pielke Sr., is a professor in the Department of Atmospheric Science at Colorado State University. He is also state climatologist for Colorado and was president of the American Association of State Climatologists from 2002 to 2003. His research areas include the study of global, regional, and local weather and climate phenomena through the use of sophisticated mathematical simulation models and observational datasets. He has published widely on the role that land-use change and vegetation dynamics may play as a driver of observed changes in climate. He has served as chairman and member of the American Meteorological Society (AMS) Committee on Weather Forecasting and Analysis, and was chief editor for the *Monthly Weather Review* from 1981 to 1985 and co-chief editor of the *Journal of Atmospheric Science* from 1995 to 2000. He was elected a fellow of the AMS in 1982. Dr. Pielke previously served on the NRC's Committee on Carbon Monoxide Episodes in Meteorological and Topographical Problem Areas and Panel on Coastal Meteorology. Dr. Pielke received a Ph.D. in meteorology from Pennsylvania State University.

Dr. Veerabhadran Ramanathan is a professor and director at the Center for Atmospheric Sciences and the Center for Clouds, Chemistry and Climate at the University of California, San Diego's Scripps Institution of Oceanography. Through his research, Dr. Ramanathan has identified chlorofluorocarbons, stratospheric ozone, and tropospheric aerosols as significant factors in anthropogenic climate change. As principal investigator for the NASA Radiation Budget Experiment, he demonstrated that clouds had a global radiative cooling effect. He was the co-chief scientist for the Indian Ocean Experiment (INDOEX), which led to the discovery of widespread atmospheric brown clouds over the Indian Ocean and South Asia. Dr. Ramanathan was inducted into the National Academy of Sciences in 2002 and is a fellow of the American Geophysical Union, the American Meteorological Society, and the American Association for the Advancement of Science. He earned his Ph.D. in planetary atmospheres from the State University of New York at Stony Brook. He has previously served on the NRC Board on Global Change and Climate Research Committee.

Dr. Lynn M. Russell is an associate professor in the Center for Atmospheric Sciences, Scripps Institution of Oceanography, University of California, San Diego. Her research is in the area of aerosol particle chemistry, including the behavior of particles under pristine and anthropogenically influenced conditions. Her research interests span experimental and modeling approaches to aerosol evolution in the atmosphere, incorporating chemical and physical mechanisms in aerosol-cloud interactions, organic aerosols, and their radiative effects. She has served on several NRC committees, including the Panel on Aerosol Radiative Forcing and Climate Change, the Committee to Review NARSTO's Scientific Assessment of Airborne Particulate Matter, and the Panel on Atmospheric Effects of Aviation. She holds a Ph.D. in chemical engineering from the California Institute of Technology.

NRC STAFF

Dr. Amanda C. Staudt is a senior program officer with the Board on Atmospheric Sciences and Climate of the National Academies. She received an A.B. in environmental engineering and sciences and a Ph.D. in atmospheric sciences from Harvard University. Her doctorate research involved developing a global three-dimensional chemical transport model to investigate how long-range transport of continental pollutants affects the chemical composition of the remote tropical Pacific troposphere. Since joining the National Academies in 2001, Dr. Staudt has staffed the National Academies review of the U.S. Climate Change Science Program Strategic Plan and the long-standing Climate Research Committee. Dr. Staudt has also worked on studies addressing air quality management in the United States, research priorities for airborne particulate matter, the NARSTO Assessment of the Atmospheric Science on Particulate Matter, weather research for surface transportation, and weather forecasting for aviation traffic flow management.

Dr. Parikhit Sinha is a program officer with the Board on Atmospheric Sciences and Climate of the National Academies. He received an A.B. in environmental engineering and sciences from Harvard University and a

Ph.D. in atmospheric sciences from the University of Washington, Seattle. His doctorate research involved airborne measurements and chemical transport modeling of trace gas and particle emissions from savanna fires in southern Africa. Since joining the National Academies in 2004, Dr. Sinha has worked on studies addressing climate change indicators, estimating and communicating uncertainty in weather forecasts, and climate variability and change in Asia.

Ms. Elizabeth A. Galinis is a senior program assistant for the Board on Atmospheric Sciences and Climate. She received her B.S. in marine science from the University of South Carolina in 2001. Since her start at the National Academies in March 2002, she has worked on studies involving next-generation weather radar (NEXRAD), weather modification, climate sensitivity, and climate change. Ms. Galinis is pursuing a master's degree in environmental science and policy at Johns Hopkins University.

B

Statement of Task

This study will examine the current state of knowledge regarding the direct and indirect radiative forcing effects of gases, aerosols, land-use, and solar variability on the climate of the Earth's surface and atmosphere and it will identify research needed to improve our understanding of these effects. Specifically, this study will:

1. Summarize what is known about the direct and indirect radiative effects caused by individual forcing agents, including the spatial and temporal scales over which specific forcing agents may be important;

2. Evaluate techniques (e.g., modeling, laboratory, observations, and field experiments) used to estimate direct and indirect radiative effects of specific forcing agents;

3. Identify key gaps in the understanding of radiative forcing effects on climate;

4. Identify key uncertainties in projections of future radiative forcing effects on climate;

5. Recommend near- and longer-term research priorities for improving our understanding and projections of radiative forcing effects on climate.

C

Glossary and Acronyms

AAO   Antarctic Oscillation

Abrupt climate change   An abrupt climate change occurs when the climate system is forced to cross some threshold, triggering a transition to a new state at a rate determined by the climate system itself and faster than the cause. (NRC, 2002)

ACE 1 and 2   Aerosol Characterization Experiments

ACRIM   Active Cavity Radiometer Irradiance Monitor

Aerosol   A colloidal system in which the dispersed phase is composed of either solid or liquid particles and in which the dispersion medium is some gas, usually air. There is no clear-cut upper limit to the size of particles composing the dispersed phase in an aerosol, but as in all other colloidal systems, it is rather commonly set at 1 μm. Haze, most smokes, and some fogs and clouds may thus be regarded as aerosols. However, it is not good usage to apply the term to ordinary clouds with drops so large as to rule out the usual concept of colloidal stability. It is also poor usage to apply the term to the dispersed particles alone; an aerosol is a system of dispersed phase and dispersing medium taken together. (American Meteorological Society [AMS])

AIM   Asia-Pacific Integrated Model

Albedo   The ratio of reflected flux density to incident flux density, referenced to some surface. Albedos commonly tend to be broadband ratios, usually referring either to the entire spectrum of solar radiation or to just the visible portion. More precise work requires the use of spectral albedos, referenced to specific wavelengths. Visible albedos of natural surfaces range from low values of 0.04 for calm, deep water and overhead Sun, to 0.8 for fresh snow or thick clouds. Many surfaces show an increase in albedo with increasing solar zenith angle. (AMS)

ALE   Atmospheric Lifetime Experiment

AMIP   Atmospheric Model Intercomparison Project

Amphiphilic   Of, relating to, or being a compound (as a surfactant) consisting of molecules having a polar, water-soluble group attached to a water-insoluble hydrocarbon chain; *also*: being a molecule of such a compound.

AO   Arctic Oscillation

AOGCM   Atmosphere-Ocean General Circulation Model

ARGO   A global array that will eventually include approximately 3000 free-drifting profiling floats that measure the temperature and salinity of the upper 2000 m of the ocean.

ARM   Atmospheric Radiation Measurement

ASF Model   Atmospheric Stabilization Framework Model

A-Train   A planned satellite formation consisting of two of the major EOS missions, three Earth System Science Pathfinder missions, and a French Centre National d'Etudes Spatiales (CNES) mission flying along the same orbit track and separated by only a few minutes.

AVHRR   Advanced Very High Resolution Radiometer

Black carbon (BC)   Light-absorbing carbonaceous aerosol, including elemental carbon and low-volatility organic compounds.

BLAG (Berner/Lasaga/Garrels) hypothesis   Variations in seafloor spreading rates lead to variations in volcanic outgassing and, thus, atmospheric CO_2 concentrations.

Bowen ratio   The ratio of sensible to latent heat fluxes from the Earth's surface up into the atmosphere. Typical values are 5 over semiarid regions, 0.5 over grasslands and forests, 0.2 over irrigated orchards or grass, 0.1 over the sea, and negative in some advective situations such as over oases where sensible heat flux can be downward while latent heat flux is upward. (AMS)

BSRN   Baseline Surface Radiation Network

CCN   Cloud condensation nuclei

CFC   Chlorofluorocarbon

Climate feedback   An amplification or dampening of the climate response to a specific forcing due to changes in the atmosphere, oceans, land, or continental glaciers.

Climate forcing   An energy imbalance imposed on the climate system either externally or by human activities.

Climate model   A simplified mathematical representation of the Earth's climate system.

Climate response   Change in the climate system resulting from a climate forcing.

Climate sensitivity parameter or climate feedback parameter (λ)   The equilibrium global mean temperature change (°C) for a 1 W m^{-2} TOA radiative forcing. λ is typically in the range of 0.3-1.4°C m^2 W^{-1} in the current generation of GCMs. Climate sensitivity has played a central role in interpretation of model outputs, in evaluation of future climate changes expected from various scenarios, and it is closely linked to attribution of currently observed climate changes. An ongoing challenge to models and to climate projections has been to better define this key parameter and to understand the differences in computed values between various models.

Climate system   The system consisting of the atmosphere, hydrosphere, lithosphere, and biosphere, determining the Earth's climate as the result of mutual interactions and responses to external influences (forcing). Physical, chemical, and biological processes are involved in interactions among the components of the climate system. (AMS)

Coupled Model   A class of analytical or numerical time-dependent models in which at least two different subsystems of Earth's climate system are allowed to interact. These subsystems may include the atmosphere, hydrosphere, cryosphere, and biosphere. This term is most commonly used for models of the evolution and interaction of Earth's atmosphere and ocean. Coupled (two-way) interaction between different subsystems can be contrasted with the class of models in which the evolution of subsystem A is affected by the present state of subsystem B, but changes in A do not feed back on the evolution of B itself. (AMS)

CMDL   Climate Monitoring and Diagnostics Laboratory

Cryosphere   That portion of the Earth where natural materials (water, soil, etc.) occur in frozen form. Generally limited to the polar latitudes and higher elevations. (AMS)

CSM   Climate System Model

CTM   Chemical transport model

Direct forcing   Climate forcing that directly effects the radiative budget of the Earth's climate system. For example, this perturbation may be due to a change in concentration of the radiatively active gases, a change in solar radiation reaching the Earth, or changes in surface albedo. Radiative forcing is reported in the climate change scientific literature as a change in energy flux at the tropopause, calculated in units of watts per square meter (W m^{-2}); model calculations typically report values in which the stratosphere was allowed to adjust thermally to the forcing under an assumption of fixed stratospheric dynamics.

DSCOVR   Deep Space Climate Observatory

DVI   Dust veil index

EBM   Energy Balance Model

ECHAM   GCM based on European Centre for Medium-Range Weather Forecasting forecast models, modified and extended in Hamburg

ECMWF   European Centre for Medium-Range Weather Forecasts

EDGAR   Emissions Database for Global Atmospheric Research

Efficacy   The ratio of the climate sensitivity parameter λ for a given forcing agent to λ for a doubling of CO_2 ($E = \lambda/\lambda_{CO_2}$). The efficacy E is then used to define an effective forcing $F_e = f E$.

ENSO   El Niño/Southern Oscillation

EOS   Earth Observing System

ERBE   Earth Radiation Budget Experiment

ERBS   Earth Radiation Budget Satellite

Evapotranspiration   The combined processes, including physical evaporation and transpiration, through which water is transferred to the atmosphere from open water and ice surfaces, bare soil, and vegetation that make up the Earth's surface. Over bare soils or the ocean, only physical evaporation occurs.

FN   Freezing nuclei

GCM   General circulation model

GCR   Galactic cosmic rays

GDP   Gross domestic product

General circulation model A time-dependent numerical model of the atmosphere. The governing equations are the conservation laws of physics expressed in finite-difference form, spectral form, or finite-element form. Evolution of the model circulation is computed by time integration of these equations starting from an initial condition. A GCM can be used for weather prediction or for climate studies. (AMS)

Geological timescale Time as considered in terms of the history of the Earth. It is divided into geologic eras, periods, and epochs. Depending on the part of the geologic timescale, increments are as long as tens of millions of years or as short as hundreds of years. In general, geologic time is more finely divided closer to the present. (AMS)

Global warming potential (GWP) An index describing the radiative characteristics of well-mixed greenhouse gases that represents the combined effect of the differing times these gases remain in the atmosphere and their relative effectiveness in absorbing outgoing infrared radiation. This index approximates the time-integrated warming effect of a unit mass of a given greenhouse gas in today's atmosphere, relative to that of carbon dioxide.

GPS Global positioning system

Greenhouse gases Those gases, such as water vapor, carbon dioxide, ozone, methane, nitrous oxide, and chlorofluorocarbons, that are fairly transparent to the short wavelengths of solar radiation but efficient at absorbing the longer wavelengths of infrared radiation emitted by the Earth and atmosphere. Trapping of heat by these gases controls the Earth's surface temperature, despite their presence in only trace concentrations in the atmosphere. Anthropogenic emissions are important additional sources for all except water vapor. Water vapor, the most important greenhouse gas, is thought to increase in concentration in response to increased concentrations of the other greenhouse gases as a result of feedbacks in the climate system. (AMS)

GRIP/GISP Greenland Ice Core Project/Greenland Ice Sheet Project

Holocene epoch The last 10,000 years of geologic time.

Hygroscopicity The relative ability of a substance (as an aerosol) to adsorb water vapor from its surroundings and ultimately dissolve. (AMS)

IGACO Intergrated Global Atmospheric Chemistry Observation System

IIASA International Institute for Applied Systems Analysis

IMAGE Integrated Model to Assess the Greenhouse Effect

Indirect radiative forcing A climate forcing that creates a radiative imbalance by first altering climate system components (e.g., precipitation efficiency of clouds), which then almost immediately lead to changes in radiative fluxes. Examples include the effect of solar variability on stratospheric ozone and the modification of cloud properties by aerosols.

INDOEX   Indian Ocean Experiment

Infrared radiation   That portion of the electromagnetic spectrum lying between visible light and microwaves. The wavelength range is approximately between 720 and 1 μm. In meteorology, this range is often further divided into the solar infrared and terrestrial radiation, with the division occurring around 4 μm. Dominant absorbers of infrared radiation include the Earth's surface, clouds, water vapor, and carbon dioxide. According to Kirchhoff's law, these are also good emitters of infrared radiation. (AMS)

IPAT   The $I = P \times A \times T$ framework hypothesizes that environmental impact (I) is determined by the interacting effects of population size (P), per capita consumption levels (A, for affluence), and finally the per capita pollution generated by the technology (T) used to satisfy the consumption levels.

IPCC   Intergovernmental Panel on Climate Change

IR   Infrared

ISCCP   International Satellite Cloud Climatology Project

LAI   Leaf area index

Latent heat   The specific enthalpy difference between two phases of a substance at the same temperature. The latent heat of vaporization is the water vapor-specific enthalpy minus the liquid water-specific enthalpy. When the temperature of a system of dry air and water vapor is lowered to the dewpoint and water vapor condenses, enthalpy released by the vapor heats the air vapor liquid system, reducing or eliminating the rate of temperature reduction. Similarly, when liquid water evaporates, the system must provide enthalpy to the vapor by cooling. The latent heat of fusion is the specific enthalpy of water minus that of ice, and the latent heat of sublimation is the specific enthalpy of water vapor minus that of ice. (AMS)

LBA   Large-scale Biosphere-Atmosphere Experiment in Amazonia

LIP   Large igneous province

Longwave radiation   In meteorology, a term used loosely to distinguish radiation at wavelengths longer than about 4 μ, usually of terrestrial origin, from that at shorter wavelengths (shortwave radiation), usually of solar origin. (AMS)

MARIA Multiregional Approach Resource and Industry Allocation

MESSAGE Model for Energy Supply Strategy Alternatives and General Environmental Impact

MiniCAM Mini Climate Assessment Model

MODIS Moderate Resolution Imaging Spectroradiometer

MSU Microwave Sounding Unit

MWR Microwave Radiometer

NACP North American Carbon Program

NAM Northern Annular Mode

NAO North Atlantic Oscillation

NASA GISS National Aeronautics and Space Administration Goddard Institute for Space Studies

NCEP National Centers for Environmental Prediction

NIST National Institute of Standards and Technology

NMVOC Non-methane volatile organic compounds

Nonradiative forcing A climate forcing that creates an energy imbalance that does not immediately involve radiation. An example is the increasing flux resulting from agricultural irrigation.

NPOESS National Polar-orbiting Operational Environmental Satellite System

NRC National Research Council

NWP Numerical Weather Prediction

OCO Orbiting Carbon Observatory

OECD Organization for Economic Cooperation and Development

Optical depth The optical thickness measured vertically above some given altitude. Optical depth is dimensionless and may be used to specify many different radiative characteristics of the atmosphere. (AMS)

PBL Planetary boundary layer

PCM Parallel Climate Model

PETM Paleocene-Eocene Thermal Maximum

POLDER Polarization and Directionality of the Earth's Reflectances

Projections of climate change   An estimate of future climate, typically produced by a climate model, in response to estimates of future natural and
anthropogenic forcings. Note that most projections consider only a subset
of possible forcings.

Proxy data   Data gathered from natural recorders of climate variability
(e.g., tree rings, ice cores, fossil pollen, ocean sediments, coral and historical
data). By analyzing records taken from these and other proxy sources,
scientists can extend understanding of climate far beyond the 140-year
instrumental record.

PSC   Polar stratospheric cloud

QBO   Quasi-Biennial Oscillation

Radiative forcing   See *direct radiative forcing; indirect radiative forcing*

RCCP   Regional climate change potential

RCE   Radiative-convective equilibrium

Sensible heat   The outcome of heating a surface without evaporating water from it. Sensible heat per unit mass can be identified roughly with the
specific enthalpy of unsaturated air, that is, approximately $c_{pd}T$, where c_{pd}
is the specific heat of dry air at constant pressure and T is temperature.
Sensible heat is often compared with latent heat, which is the difference
between the enthalpy of water vapor and that of liquid water. (AMS)

Shortwave radiation   In meteorology, a term used loosely to distinguish
radiation in the visible and near-visible portions of the electromagnetic
spectrum (roughly 0.4 to 4.0 μm in wavelength), usually of solar origin,
from that at longer wavelengths (see *longwave radiation*), usually of terrestrial origin. (AMS)

SIRCUS   Spectral Irradiance and Radiance Calibrations with Uniform
Sources

Solar insolation   The amount of electromagnetic energy (solar radiation) incident on the surface of the Earth, generally expressed in
kW h m^{-2} day^{-1}. (AMS)

Solar irradiance   The amount of solar energy that arrives at a specific
area of a surface during a specific time interval (radiant flux density). A
typical unit is W m^{-2}.

SORCE   Solar Radiation and Climate Experiment

SRES   Special Report on Emissions Scenarios produced by the IPCC.
(Nakićenović, 2000)

SST   Sea surface temperature

Stratosphere   The region of the atmosphere extending from the top of the troposphere (the tropopause, located at roughly 10-17 km above the surface) to the base of the mesosphere (the stratopause, located at roughly 50 km above the surface). The stratosphere is characterized by constant or increasing temperatures with increasing height and marked vertical stability. It owes its existence to heating of ozone by solar UV radiation, and its temperature varies from 85°C or less near the tropical tropopause to roughly 0°C at the stratopause. Although the major constituents of the stratosphere are molecular nitrogen and oxygen, just as in the troposphere, the stratosphere contains a number of minor chemical species that result from photochemical reactions in the intense ultraviolet radiation environment. Chief among these is ozone, whose presence shelters the underlying atmosphere and the Earth's surface from exposure to potentially dangerous UV radiation. (AMS)

TAR   IPCC Third Assessment Report

TARFOX   Tropospheric Aerosol Radiative Forcing Observational Experiment

Teleconnection   (1) A linkage between weather changes occurring in widely separated regions of the globe. (2) A significant positive or negative correlation in the fluctuations of a field at widely separated points. Most commonly applied to variability on monthly and longer timescales, the name refers to the fact that such correlations suggest that information is propagating between the distant points through the atmosphere. (AMS)

TES   Tropospheric Emission Spectrometer

THC   Thermohaline circulation

TOA   Top of the atmosphere

TOMS   Total Ozone Mapping Spectrometer

TOPEX   Topography Experiment for Ocean Circulation

Tropopause   The boundary between the troposphere and stratosphere, usually characterized by an abrupt change of lapse rate. The change is in the direction of increased atmospheric stability from regions below to regions above the tropopause. Its height varies from 15 to 20 km (9 to 12 miles) in the Tropics to about 10 km (6 miles) in polar regions. (AMS)

Troposphere   That portion of the atmosphere from the Earth's surface to the tropopause—that is, the lowest 10-20 km (6-12 miles) of the atmosphere—and the portion of the atmosphere at which most weather occurs.

The troposphere is characterized by decreasing temperature with height, appreciable vertical wind motion, appreciable water vapor, and weather. (AMS)

TSI Total solar irradiance

Twomey effect The increase in cloud albedo due to an increase in aerosol concentration. For a dynamic forcing that creates a cloud with a given vertical extent and liquid water content, an increase in aerosol concentration going into the cloud can result in the formation of a larger number of smaller droplets compared to an unperturbed cloud. The end result is an increase in cloud albedo.

UARS Upper Atmosphere Research Satellite

UV Ultraviolet

VEI Volcanic Explosivity Index

Velocity potential A scalar function with its gradient equal to the velocity vector of an irrotational flow.

VOC Volatile organic compound